T0223817

An Introduction to
Models of Online Peer-to-Peer
Social Networking

Synthesis Lectures on Communication Networks

Editor

Jean Walrand, *University of California, Berkeley*

Synthesis Lectures on Communication Networks is an ongoing series of 50- to 100-page publications on topics on the design, implementation, and management of communication networks. Each lecture is a self-contained presentation of one topic by a leading expert. The topics range from algorithms to hardware implementations and cover a broad spectrum of issues from security to multiple-access protocols. The series addresses technologies from sensor networks to reconfigurable optical networks. The series is designed to: Provide the best available presentations of important aspects of communication networks; Help engineers and advanced students keep up with recent developments in a rapidly evolving technology; Facilitate the development of courses in this field.

An Introduction to Models of Online Peer-to-Peer Social Networking

George Kesidis

ISBN: 978-3-031-79997-6   paperback
ISBN: 978-3-031-79998-3   ebook

DOI 10.1007/978-3-031-79998-3

A Publication in the Springer series
SYNTHESIS LECTURES ON COMMUNICATION NETWORKS

Lecture #8
Series Editor: Jean Walrand, *University of California, Berkeley*
Series ISSN
Synthesis Lectures on Communication Networks
Print 1935-4185   Electronic 1935-4193

An Introduction to Models of Online Peer-to-Peer Social Networking

George Kesidis
Pennsylvania State University

SYNTHESIS LECTURES ON COMMUNICATION NETWORKS #8

ABSTRACT

This book concerns peer-to-peer applications and mechanisms operating on the Internet, particularly those that are not fully automated and involve significant human interaction. So, the realm of interest is the intersection of distributed systems and online social networking. Generally, simple models are described to clarify the ideas. Beginning with short overviews of caching, graph theory and game theory, we cover the basic ideas of structured and unstructured search. We then describe a simple framework for reputations and for iterated referrals and consensus. This framework is applied to a problem of sybil identity management. The fundamental result for iterated Byzantine consensus for a relatively important issue is also given. Finally, a straight-forward epidemic model is used to describe the propagation of malware on-line and for BitTorrent-style file-sharing.

This short book can be used as a preliminary orientation to this subject matter. References are given for the interested student to papers with good survey and tutorial content and to those with more advanced treatments of specific topics. For an instructor, this book is suitable for a one-semester seminar course. Alternatively, it could be the framework for a semester's worth of lectures where the instructor would supplement each chapter with additional lectures on related or more advanced subject matter. A basic background is required in the areas of computer networking, probability theory, stochastic processes, and queueing.

KEYWORDS

computer and communication networking, social networking, peer-to-peer networking, reputations, referrals, consensus, epidemics

For Selena, Emmaline, and Cleoniki.

Contents

Preface

In this book, we consider distributed applications and mechanisms operating on the Internet, particularly those that are not fully automated and involve significant human interaction. There is no focus here on wireless, including mobile, networking issues. Thus, our realm of interest is the intersection of distributed online peer-to-peer and social networking. Generally, simple models are developed to clarify the ideas. For example, simple game-theoretic models of end-user behavior are described assuming synchronized player action and idealized networking dynamics.

This short book is intended as a preliminary orientation for a one-semester seminar course. Alternatively, it could form a framework for a semester's worth of lectures where the instructor would supplement each chapter with additional lectures on related subject matter or give more advanced treatment than that given here. References to papers with good survey and tutorial content are given for this purpose.

The outline of this book is as follows. We first give an overview of networking, graphs, and games. We then discuss search in both structured and unstructured settings. The third part of this book is concerned with reputations, referrals, and file-sharing. We conclude with material on consensus and epidemics.

The requirements of the book are undergraduate-level courses in:

- communication/computer networking based on, e.g., [91, 124];

- probability theory based on, e.g., [52];

- stochastic processes based on, e.g., [67, 79], and queueing based on, e.g., [157].

This work was supported by the National Science Foundation of the United States under grants 0524202 and 0916176. I also wish to acknowledge the support of the Newton Institute of Cambridge University, Cisco Systems, and the J⁴W family for allowing me to house-sit while writing. I wish to thank my colleagues, collaborators, and students for many helpful discussions, particularly Stephane Caron, Chris Griffin and Takis Konstantopoulos. Also, I thank my wife Diane for a final proofread. Please notify the author of any errors found in this book, all of which are his responsibility.

George Kesidis
State College, Pennsylvania
kesidis@gmail.com
November 2010

PART I

Background

CHAPTER 1

Networking overview

In this chapter, we give a high-level overview of the Internet and certain distributed applications that run on it, including those used to mount social networks. Much of what is discussed below is fleshed out in greater detail in RFCs and Internet Drafts freely available (and searchable) at URL www.ietf.org, and in standard texts such as current editions of [91, 124]. Some of the principles discussed in the latter sections of this chapter will be echoed in subsequent chapters.

1.1 THE INTERNET

The Internet is a vast distributed and diverse communication network largely based on the Internet Protocol version 4 (IPv4) at present[1]. Many different types of applications operate over the Internet, particularly over client-server world-wide web (WWW) and peer-to-peer (p2p) overlays. In particular, generic social networking frameworks, now typically mounted on particular web sites, can also operate p2p. In turn, more specific applications (e.g., games, file-sharing and messaging) can be mounted on generic social networking frameworks. Thus, there is a hierarchy or layering of applications that can run over the Internet.

Also, IPv4 itself lives in a hierarchy of communication protocols at layer number 3. Layer 4, the transport layer, is used by end-hosts to interface aforementioned applications to IP. Messages are sometimes segmented and encapsulated into different packet/frame formats as they pass down across such layers (decapsulated and reassembled as they pass up). In particular, IPv4 packets are sometimes encapsulated into Ethernet frames for transport, where Ethernet is a "wired" communication protocol operating in layer 2. Ethernets are typically found in local area networks at the periphery of the Internet or on point-to-point links between Internet routers. Wireless and optical communication layer 2 networks are also found at the periphery of the Internet. Physical electromagnetic signals which represent the bits that form Ethernet frames (and the IPv4 packets which the frames encapsulate) reside in layer 1.

ADDRESSING IN LAYER 3

Under IPv4, 32-bit addresses are assigned to the computers that communicate over the Internet. Data packets that traverse the Internet are inscribed with 32-bit addresses, one each for the sending computer (source address) and receiving computer (destination address). The Internet accommodates a multicasting ability involving "group" destination addresses, as discussed below.

---

[1]The deployment of IPv6 (with a larger address space in particular) remains relatively small, although it has grown throughout the world over the last decade.

Consider two different IPv4 networks, one the public Internet familiar to us in the West, the other an independently administered network (with a separate set of 2^{32} addresses) operating in China. Network address translation (NAT) routers at the boundary translate destination addresses and (overloaded) destination port numbers on packets arriving to the router from the "Western" domain, to the IPv4 destination address and port number for the intended recipient Chinese end-host. The mapping is reversed for packets outbound from the Chinese domain[2]. NATs have enabled greater penetration of IPv4 throughout the world in at least two ways. NATs have effectively extended the address space of public Internet as we've just mentioned (originally, the assigned number of IPv4 addresses for China matched that of Stanford University). Also, a separate IPv4 domain can be administered in a way more consistent with the local socio-political aims and customs.

Note that private-enterprise and government IPv4 domains and subnets are managed very differently than public ones in the West, e.g., many FCC regulations, including application neutrality, apply only to the public IPv4 domain.

NAME RESOLUTION

Since the great majority of human beings are not aware of their specific numerical values, the desired 32-bit destination IP addresses of packets need to be automatically discovered beginning with names that human beings do understand for a web server, e-mail server, etc. For example, consider an e-mail sent to inbox@partyu.edu to inquire about a fraternity. The packets constituting that e-mail travel across the Internet with destination 32-bit address that corresponds to the mail server named mail.partyu.edu. This 32-bit destination address is automatically discovered by the Internet's domain-name system (DNS) of hierarchical name-resolution servers, where an "authoritative" name-server at Party University is ultimately responsible to resolve the name mail.partyu.edu. Once discovered, the correspondence between the name mail.partyu.edu and, for future convenience, its 32-bit equivalent may be *cached* elsewhere in DNS system and in communicating end-host computers (particularly the one that dispatched the e-mail in this example), i.e., cached to reduce the workload on the DNS, *cf.* Sec. 1.3.

PACKETIZATION

A fundamental design feature of the Internet is that messages are packetized. The sending end-host will need to segment a long message (greater than 1500 bytes) for transmission over the Internet and the receiving end-host will need to reassemble the message.

We briefly mention some transmission latency issues associated with packetization. First, consider a real-time application, e.g., Voice over IP (VoIP) with data generated at about 8 kbps (8000 bits per second). So, it takes about 0.1 s to fill a packet whose payload is 100 bytes and thereby a *packetization delay* of about 0.1 seconds is suffered by the first bit arriving to the payload. This is about the most that can be tolerated before degradation in voice quality begins to be perceived by

---

[2]On an obviously much smaller scale, NAT technology is also used in the wireless router for Internet access of multiple at-home computers using one residential broadband subscription.

the conversing humans. The IP packet header is large, on the order of 40 bytes, so a much smaller payload size would lead to reduced transmission efficiency (ratio of total payload to total payload-plus-header sizes of a packet stream) and results in larger per-packet processing overhead. Second, intermediate switches and routers need to completely receive a packet before they can relay it. Thus, a delay of at least that required to *store and forward* the packet is incurred at each such device on the packet's path from sending end-host to receiving end-host. This delay is naturally proportional to the size of the packet too; see Sec. 8.2 on store-and-forward economies of message segmentation. In summary, smaller packets tend to result in less store-and-forward and packetization delays but more transmission overhead.

ETHERNET IN LAYER 2

Peripheral Ethernet local area networks (LANs) operate in layer-2 using 48-bit Medium Access Control (MAC) addresses. Ethernet frames have MAC addresses that locally correspond to the IP addresses of the IP packet they encapsulate, while IP addresses of end-hosts located outside of the LAN are typically mapped to interfaces of the gateway routers.

These layer 2 to 3 address correspondences are determined by Ethernet's Address Resolution Protocol (ARP). Ethernet MAC addresses do not have a hierarchical structure, i.e., the structure is "flat" so that MAC addresses have no relation to physical location. Ethernet operates using a less scalable networking framework than IP inter-domain.

UNICAST FORWARDING AND ROUTING

In the Internet, network routers continually run routing protocols to decide how to forward packets according to their destination IP addresses. IPv4 is "connectionless" in the sense that layer-3 paths are not expressly set up prior to packet transmission. The Internet is composed of different domains that are used for routing (and addressing) purposes.

Within a domain, a common routing protocol is the *link-state* OSPF which employs periodic flooding to convey the local (intra-domain) topology and its link states to all participating routers. Link states are Boolean quantities $\in \{1, \infty\}$ indicating link availability and absence of significant congestion (1) or otherwise (∞). The lengths of routes/paths, given in "hops", are equal to the number of links constituting them.

BGP is a path-vector protocol that operates on a grand inter-domain scale among self-managed networks called Autonomous Systems (ASes). BGP path-vectors are only exchanged by neighboring BGP speakers as a flooding strategy would unscalably involve too much messaging. Also, the addressing system of IP is spatially hierarchical allowing for effective forwarding based only on address prefixes and, thus, greatly reducing the required sizes of the forwarding tables in Internet routers and the sizes of BGP messages. Path vectors are used because the geodesic (in hops) path lengths are typically not very long in the Internet, so the additional messaging overhead remains feasible. With path vectors the carriers can easily avoid loops in their inter-AS routes and implement forwarding policies (policies often result in sub-optimal path length by avoiding costly, from the forwarding carrier's point-of-view, network-to-network boundaries).

Finally, it's important to mention that connection-oriented Multiprotocol Label Switching (MPLS) technology has a significant presence in the Internet, particularly for business-to-business communication and in the form of permanent "virtual paths" between certain ASes within a single carrier's domain [160].

MULTICAST

The Inter-Group Messaging Protocol (IGMP) provides a scalable multicasting framework for the network layer 3. A block of IPv4 Class D addresses, signified by a special 4-bit prefix, is reserved for IGMP. This said, a layer-3 approach to unidirectional multicast or bidirectional conferencing is not warranted for smaller-scale multicasting sessions for which peer-to-peer solutions may be more appropriate. In the United States, transmission to the residential head-ends of television content is performed by private IP (IPTV) networks running IGMP.

More abstractly, a multicast framework for N participants seeks to reduce the communication costs associated with N (one sender and $N - 1$ subscribers) or $\binom{N}{2}$ (an N party conference) separate unicast sessions. To this end, individual IP routers are capable of performing parallel "copying" operations on multicast packets. The result is that multicast packets arriving at one interface can simultaneous depart on multiple other interfaces, so that a router can act as a interior vertex in a multicast tree. Ethernet switches also possess such a capability for layer-2 multicast or broadcast operations.

To avoid the complexities of setting up and maintaining optimal spanning trees end-to-end, a dynamic prune/graft framework can be used involving both routers and end-hosts. A host H wishing to join into layer-3 multicast session X transmits a subscribe message indicating the X's class-D address. Upon receipt of this message, H's gateway router R typically takes one of two possible actions:

- If R is already tapping into X on behalf of another of its downstream clients, then R adds H to the subscription list for X and simply starts copying and transmitting packets of X with destination H.

- Else, R needs to find a proximal router S which is forwarding packets of X and tap into X on behalf of H.

In the second case, R may find such a router S through a limited-scope flood of a query message for X; each query response may possess information about the length of the querying path so that the minimum path-length join operation can be determined by R. Through observation time-out of subscription indicators or receipt of explicit leave messages, R can determine that none of its hosts are presently subscribing to X and may then dispatch a prune message to its tributary router S so as to stop the flow of X packets.

For the example of IPTV, the popular commercial TV channels are typically broadcast (pushed-out) to all residential head-ends. The head-ends would more dynamically subscribe to pay-per-view video content (pulled-in on demand).

TRANSMISSION CONTROL IN LAYER 4

The role of Layer 4 mechanisms is to interface between the higher application layers of the end-hosts and the IP communication layer 3. Layer 4 is responsible for outbound message segmentation and packetization/encapsulation, and inbound message decapsulation and reassembly. Operating in layer 4, UDP is a connectionless and light-weight though unreliable protocol, whereas TCP is reliable and connection-oriented. However, TCP's method for recovering packets deemed lost in transit is too slow for interactive real-time and streaming media applications, which often run standard (e.g., RTP/RTCP) or customized error recovery and flow-control mechanisms over UDP.

TCP also employs an additive increase and multiplicative decrease (AIMD) rule to alleviate network congestion. AIMD employs first-order autoregressive estimates of session round-trip times (RTTs) to define (and respond to) packet acknowledgement time-outs. Unfortunately, TCP congestion control is easily undermined either by combining it in the network with traffic running over UDP (traffic that is nominally not as responsive to congestion) or by "selfish" end-users opting out of the communal TCP congestion control mechanism. For a recent example, UDP clients for the peer-to-peer file-sharing system BitTorrent have been introduced [145].

For those TCP sessions that cooperatively participate in network congestion control, the "fairness" of AIMD has been argued as follows. Suppose two identical sessions[3] are sharing a bottleneck bandwidth resource in the network of amount B, say measured in Mbps. Their initial transmission rates (x_1, x_2) are indicated by a black dot in Fig. 1.1 with $x_2 > x_1$, i.e., an initially "unfair" scenario. In the absence of congestion, i.e., $x_1 + x_2 \leq B$, both rates increase additively and at the same rate, i.e., parallel to the line $x_1 = x_2$. Once congested, i.e., $x_1 + x_2 > B$, the effects of excessive delay or packet loss will register with the end-users and they will exponentially back-off, i.e., $(x_1, x_2) \to (x_1/2, x_2/2)$: half the distance to the origin. Thus, congestion will abate, the cycle will repeat as a zig-zag in the figure eventually converging to the line $x_1 = x_2$, i.e., equitable throughput.

Now consider a game with n players where player i transmits at rate $\xi_i \leq \lambda$. The players share a packet transmission capacity $B < n\lambda$ bit/s, i.e., less than the peak rate of the aggregate arriving traffic. The transmission rate of a player ξ (i.e., their "play") is not necessarily their throughput, x, i.e., the payoff of the player $x \leq \xi$. Consider the vector of collective plays $\underline{\xi}$. Suppose that the payoff of player i is given by

$$x_i := \xi_i \mathbf{1}\{\sum_{j=1}^{n} \xi_j \leq B\}. \tag{1.1}$$

[3]i.e., with same RTT, responsiveness, etc.

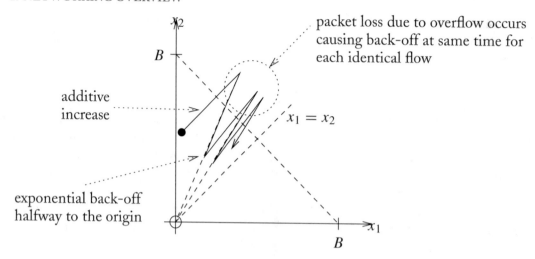

Figure 1.1: Fairness of additive increase and multiplicative decrease (AIMD).

That is, if $\sum_{j=1}^{n} \xi_j > B$ then everyone gets zero payoff because the queue is unstable[4]. Consider a feasible[5] set of collective plays $\underline{\xi}$ satisfying

$$\sum_i \xi_i \;=\; B-, \tag{1.2}$$

i.e., the aggregate arrival rate is just less than B. All such $\underline{\xi}$ are *Nash equilibrium points* (NEPs) because if any single player i unilaterally changes their play (i.e., all other players $j \neq i$ continue to play ξ_j) then i will experience reduced payoff. Note that we have not associated a usage-based cost of a play consistent with volume-unlimited Internet access for a low flat-rate monthly fee (a fee which may depend on access limits λ).

   In practice, a player considering whether to adopt TCP's AIMD communal congestion control policy *knows* that the great majority of the other players will continue to follow it, and so is clearly incentivized to opt out of the TCP AIMD coalition. There are recent proposals to protect TCP congestion control by detecting congestion and throttling back "heavy hitter" (congestion non-responsive) TCP sessions, e.g., [39, 161]. Generally, in the presence of significant UDP traffic, to protect TCP's communal congestion control strategy, one needs to separately schedule the TCP traffic. A nice explanation of TCP's dynamics on a graph is given in Sec. 5 of [82].

---

[4]Either the infinite-memory queue is unstable resulting in excessive delays, or the finite-memory queue will experience heavy packet loss due to buffer overflow.
[5]That is, $\xi_i \leq \lambda_i$ for all players i.

1.2   THE CLIENT-SERVER WORLD-WIDE WEB (WWW)

The world-wide web is an enormous distributed repository of data and services accessed through web clients (HTTP browsers) or specialized "app" programs running on the end-hosts. The browsers display web pages for the end-user. End-users surf the web by clicking on links from one page to another, selecting a previously viewed page from a bookmark list cached in their local end-host, typing in a URL, or by appealing to a search engine. Web search engines, which operate on an image of the entire web (as mined by crawling processes) play the joint role of search and name resolution for the WWW. Web search engines also provide additional support by correcting keyword spellings and offering search suggestions.

COOKIES

Cookies are files held in the end-hosts that store information about web surfing history including: past sites visited, content downloaded, commercial transactions, and time-stamped certificates indicating a prior successful password-based login. Information stored in cookies enables certain conveniences for the surfing end-user. Storing such information in the end-hosts is a scalable approach from the web servers' point-of-view. However, certain information stored in cookies raises both privacy and security concerns.

RANKING PAGES

Search results are prioritized, e.g., pages can be listed in order of the number of other pages which link to them as in Google's PageRank, i.e., a measure of the "popularity" of the page. Such measures of popularity are important for setting the price of advertising on commercial web sites. A simple iterative procedure for determining the relative popularity of web pages is as follows. For a population of N pages numerically indexed $1, 2, ..., N$, let d_i be the number of different pages which are linked-to by page i, i.e., i's out-degree. Define the $N \times N$ *stochastic* matrix $\mathbf{P}$ with entries $P_{i,j} = 1/d_i$ if i links to j, otherwise $P_{i,j} = 0$ (with $P_{i,i} = 0$ for all i). Define the popularity/rank of $\pi_i \geq 0$ of page i so that the following equations are satisfied:

$$\pi_i = \sum_j \pi_j P_{j,i} \text{ for all i } \text{ and } 1 = \sum_j \pi_j. \tag{1.3}$$

Note how the j contributes to i's popularity, but that contribution is reduced through division by the total out-degree d_j of j. In matrix form, the first set of equations is simply $\underline{\pi}^\mathrm{T} = \underline{\pi}^\mathrm{T}\mathbf{P}$. So, $\underline{\pi}$ is the invariant distribution of a discrete-time Markov chain with transition probabilities $\mathbf{P}$, i.e., a random walk on the graph formed by the N web pages as vertices and the links between them as directed edges, with time corresponding to the number of transitions to other web pages (clicked-on web links). The marginal distribution of the Markov chain at time k, $\underline{\pi}(k)$ satisfies the Kolmogorov equations $\underline{\pi}^\mathrm{T}(k) = \underline{\pi}^\mathrm{T}(k-1)\mathbf{P}$ [79], i.e., $\pi_i(k)$ is the probability that the random walk is at page i at time k. If $\mathbf{P}$ is aperiodic and irreducible then there is a unique stationary/invariant distribution $\underline{\pi}$ satisfying (1.3) such that $\lim_{k\to\infty} \underline{\pi}(k) = \underline{\pi}$ [58], *cf.* Ch. 9.

Google's PageRank considers a parameter that models how web surfers do not always select links from web pages but may select links from among their bookmarks. Suppose that a bookmark selection occurs with probability b and that the probability of specific bookmarked page selected is b/N. To this end, instead of $\mathbf{P}$, an alternative is to use the stochastic matrix $\tilde{\mathbf{P}} = (1-b)\mathbf{P} + (b/N)\mathbf{1}$, where $\mathbf{1}$ is the $N \times N$ matrix all of whose entries are 1. With $0 < b \le 1$, $\tilde{\mathbf{P}}$ will be irreducible and aperiodic irrespective of $\mathbf{P}$. But since scalable computation of $\underline{\pi}$ may rely on sparseness of non-zero entries in $\mathbf{P}$, we can retain $\mathbf{P}$ and simply adjust the rank of page i to be given by $(1-b)\pi_i + b/N$. More precisely, we adjust the iteration to the affine $\underline{\tilde{\pi}}^T(k) = (1-b)\underline{\pi}^T(k-1)\mathbf{P} + (b/N)\underline{1}^T$, where $\underline{1}$ is a column vector of 1s, which leads to the stationary distribution

$$\underline{\tilde{\pi}}^T = (b/N)\underline{1}^T[\mathbf{I} - (1-b)\mathbf{P}]^{-1},$$

where $\mathbf{I}$ is the $N \times N$ identity matrix and $\mathbf{I} - (1-b)\mathbf{P}$ is non-singular for $0 < b \le 1$ because $\mathbf{P}$ is stochastic [71].

1.3   ADAPTIVE AND DISTRIBUTED CACHING

We have already discussed the distributed domain name system (DNS) which caches layer-3 IP-address to domain-name correspondences. A basic goal of content caching is to reduce traffic which may cause congestion and result in additional transit costs and access latency. That is, if a web server is very busy or network between a group of clients and the server is congested, store a copy of the desired data-object in a web cache nearby the requesting clients so that they can access it readily.

In a distributed caching system, a component cache will communicate with its assigned clients, with the other component caches, and with the authoritative/origin servers of the data objects contained in the cache. Basic problems associated with caching are:

- Where in the network to place the cache?

- Which content to store?

- How to ensure that the stored content is up to date?

Obviously, certain content will be unlikely chosen to be cached if it:

- is of little demand to the cache's clients (a cached item may be aged-out (overwritten) from a full cache according to, e.g., a simple *least recently used* rule [129]);

- changes very frequently (i.e., is not "durable") requiring repeated updating via the origin servers; or

- is very large so as to crowd out other content.

Placement of a cache closer to origin servers of its data objects will reduce overhead associated with updating and verifying the data object, but may increase overhead associated with responding to queries from the cache's clients. Given models of demand and content durability (and a "scoring" system weighing them), one can formulate optimization problems for choice of content to cache and placement of the caches themselves [129]. One can also envision web caches that anticipate/predict demand trends of particular data objects and proactively fetch them from their origin servers.

Similar problems relate to placement of search engine servers or mirrored server farms, or the mirroring of individual server systems, to make them scale to larger client communities. That is, how to divide the client community between mirrored servers [143], where to physically situate them, and how best to ensure consistency of the same content stored in different servers.

In Ch. 4, we consider a peer-to-peer caching system with a much larger number of relatively small and unreliable caches whose cooperation may need to be incentivized and caching overhead needs to be balanced. In this section, simple models will be described for an ISP's distributed web cache, which is set-up and managed for its local clients. More generic content caches have been proposed for peer-to-peer traffic[6] [159] and for streaming media [99], the latter often much less complex than the caching frameworks currently used for commercial IPTV.

Even in a regional setting, a distributed cache may be preferable to a single cache with equivalent total memory as the former will have greater input/output (memory) bandwidth and will be more robust in the presence of failures.

Again, to determine whether a particular data object ought to be stored in a cache, a least-recently-used rule could be used to reckon demand and thereby rank the data objects. Alternatively, an autoregressive average of the inter-request time could be used. And again, data objects may be prioritized in the cache according to a combination of their estimated demand, durability, and size (impacting cost-to-fetch and cost-to-transmit). Uncached data objects are also monitored to decide which of them should replace the least important entry in the cache, possibly on a trial basis.

Consider a fixed population of n caches serving a much larger client base. A client is ostensibly assigned to one cache but a cache that cannot resolve a client's query will forward that query to another cache. This may occur multiple times for a given query until the query is resolved by a cache that possesses the requested data object, or until a time-to-live (TTL) threshold is reached and the query is dispatched to the origin server.

In the following, we give a performance result for a distributed cache in a highly idealized setting. Also, we discuss how to divide and index content in caches and how to more realistically route client queries among them, by "consistent" hashing.

PERFORMANCE OF AN IDEALIZED UNSTRUCTURED MODEL

Consider the case of a large number $n \gg 1$ of component caches allowing a single data object (particularly a popular one) to be resident in more than one cache at the same time. A cache will forward a query it cannot resolve to another cache selected uniformly at random, where such forwarding decisions are also assumed independent. Finally, suppose that if a cache does not possess

---

[6]Note that ISPs operating such caches may further expose themselves to liability for copyright infringement associated with illegal file-sharing.

a data object for which it was *once* queried (i.e., a "cache miss" occurs), then it will acquire that object from its origin server and store it for a time $1/\mu$ (otherwise irrespective of demand).

Consider a query for a particular data object and assume its overall query rate is λ from the client-base of the distributed cache. Also assume that the query is first received by a cache selected uniformly at random and that, on average in steady state, a positive fraction $\varepsilon \ll 1$ of caches possess it. So, the expected number of caches EF to which the query will be forwarded (i.e., the expected number of cache misses) is about

$$EF = \sum_{k=1}^{\min\{n,T\}} k(1 - \varepsilon)^k \varepsilon \approx \frac{1}{\varepsilon},$$

where T is the TTL and we have assumed T, n are both large enough to justify approximating with a geometric distribution. Let x be the (random) number of caches that possess the data object in steady-state, i.e., its mean $Ex = n\varepsilon$. So, a single query will result in the data object "arriving to" about n/x caches[7]. See the closed queueing system model depicted in Figure 1.2.

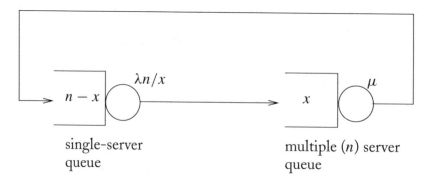

Figure 1.2: Tracking a single cached data object.

By Little's theorem [157],

$$Ex = \frac{1}{\mu}\lambda E(n/x)$$
$$\geq \frac{\lambda n}{\mu Ex},$$

where the inequality is Jensen's. Thus, $Ex \geq \sqrt{\lambda n/\mu}$. Equivalently, $\varepsilon \geq \sqrt{\frac{\lambda}{n\mu}}$ or

$$EF \leq \sqrt{\frac{n\mu}{\lambda}};$$

[7]The data object will be stored in the queried-but-missed caches.

see [130] for another formulation of this problem.

Another model illustrating a related "square-root principle" focuses on the *total* popularity of the content of the individual caches. Let Λ_i be the probability that a query can be resolved by cache i. If a cache i is not able to resolve a query, it independently decides to forward the query to the cache j with probability $\phi_j / \sum_{k \neq i} \phi_k$, i.e., ϕ_k is an attribute of cache k known to all other caches satisfying

$$\sum_k \phi_k = 1. \qquad (1.4)$$

The parameters ϕ are controllable by the caching system, whereas the parameters Λ are the inputs and not controllable by the caching system. Also, assume that each query can only be resolved by one cache, i.e., the caches have no common queried content so that $\sum_i \Lambda_i = 1$. So, the mean number of forwarding steps to resolve a query is

$$\mathsf{E}F \approx \sum_i \Lambda_i \frac{1}{\phi_i}. \qquad (1.5)$$

To minimize (1.5) subject to (1.4), we can simply define the Lagrangian

$$L(\underline{\phi}, \mu) := \sum_i \Lambda_i \frac{1}{\phi_i} + \mu(\sum_i \phi_i - 1),$$

where $\mu \geq 0$ is the Lagrange multiplier associated with constraint (1.4). Jointly minimizing L over $\underline{\phi}, \mu \geq 0$ gives that (1.4) and $\phi_i = \sqrt{\Lambda_i / \mu}$ have to hold for all i. That is, $\phi_i = \frac{\sqrt{\Lambda_i}}{\sum_j \sqrt{\Lambda_j}}$ which implies

$$\mathsf{E}F = \left(\sum_i \sqrt{\Lambda_i} \right)^2 ;$$

see [37, 100, 166] and *cf.* discussion on popularity based routing in Ch. 5. Note that, to effect this optimal forwarding rule, Λ_i (an attribute of cache i) needs to be known by all other caches $j \neq i$. Also, if $\Lambda_i = 1/n$ for all i (uniform distribution of data objects among caches), then $\mathsf{E}F = n$, so that this randomized forwarding system will, on average, perform worse than a search using a deterministic, non-repeating cache selection sequence.

With additional overhead, searching can generally be improved by maintaining a "taboo" list of previously visited caches on each query, and by using multiple query threads in parallel.

CONSISTENT (STRUCTURED) HASHING

Consistent hashing [77] provides an addressing system that "randomly" distributes the content among the component caches in such a way that:

- the number of data objects is balanced among the caches and

- readdressing/redistribution overhead is balanced and minimal in the presence of content or cache churn.

Under consistent hashing, a standard hash function h maps numerical URLs $u \in U$ of cached content to points on the circumference of a unit circle; S, i.e., $h : U \to S$. Similarly, the identities of the caches themselves are mapped to points on the circle. URLs u are assigned to the first cache x that is found clockwise from $h(u)$ on the circle; see Figure 1.3 Note that a particular data object is not assigned to multiple caches.

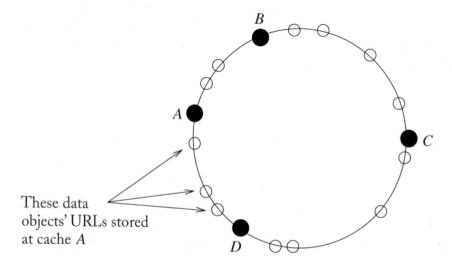

These data objects' URLs stored at cache A

Figure 1.3: Consistent hash indexing on a circle.

In [77], the forwarding performance of such a system is analyzed assuming that caches experience balanced querying loads and that the queries have a TTL after which the query is forwarded to the origin server. Moreover, after a certain number of cache misses for a certain data object u (not necessarily one, as in the idealized scenario above), a cache will decide to store u. So, there may be two types of data objects stored in a cache: those based on proximity in the hash-space (unit circle) and those based on demand.

One can add redundancy to the system by, say, proactively storing data object u at the first *and* second caches clockwise from $h(u)$. This might significantly reduce the "reactive" overhead associated with cache departures but this at the cost of additional required memory per data object. Such redundancy may be prudent in situations where caches are less reliable, as in the distributed peer-to-peer setting, *cf.* Ch. 4. Moreover, in these settings, there may be significantly more overhead[8] behind presumed knowledge of, e.g., d other caches, again in the presence of peer and content churn

---

[8]Approaches that, e.g., employ Bloom filters have been proposed to reduce this overhead, see, e.g., [90, 94].

where peer churn (arrivals and departures) is expected to be much higher than cache failure above. So, there is a need for redundancy of content to peer assignment for quicker fail-over. Uniform loading of caches is discussed in [77, 78] and in Sec. 4.3 below.

1.4 IDENTITY AUTHENTICATION

In a social network, peer nodes nominally have a unique online identity. However, in practice, peers may engage in, e.g.,

- impersonation: one peer using the identity of another without authorization;

- identity white-washing: abandoning a sullied identity and adopting a fresh new one;

- sybil identities: adopting multiple identities, e.g., to free ride, self hype, or defame another peer (*cf.* Ch. 7), and

- a shared identity: multiple peers deliberately employing the same (collective) identity.

As a result, identity disambiguation has become an important problem in online social networks. An important component of proposed solutions to this problem is an identity authentication mechanism.

To verify the putative identity of an entity in order to, e.g., deal with the threat of an impersonation attack, authentication mechanisms have employed secret passwords, pre-authorized hardware (e.g., smart cards), biometrics (e.g., fingerprints), and challenge-response (e.g., CAPTCHA human-interface verification). A common authorization framework for access to resources or data is an access control list (ACL), a very simple example of which is the file permission-bits used by operating systems such as UNIX.

In the following, we will make use of identity authentication mechanisms over unreliable channels. In a public keying system, we assume that all actors i have secure access to each others public cryptographic keys K_i. Online, this is dynamically accomplished by a reliable Key Distribution Center, or in special cases, statically prior to the actors deployment online. Each public key K_i is mated to a secret key k_i, privately held by actor i. Keys k_i and K_i are cryptographic inverses, i.e., for all messages m, the (encrypted) cypher $k_i(m)$ can be decrypted as $K_i(k_i(m)) = m$ and similarly $k_i(K_i(m)) = m$. The decryption is computationally straightforward given the mated key to that used in the encryption step[9]. Otherwise, decryption is difficult, typically at least as hard as factoring the product of two extremely large prime numbers.

If actor A wishes to authenticate her identity to actor B, A can append a digital signature σ, e.g., her identity encrypted with her private key so that $\sigma = k_A(A)$. Upon receipt, B can decrypt the putative signature $\tilde{\sigma}$ with A's public key to get $K_A(\tilde{\sigma})$, where $K_A(\tilde{\sigma}) = A$ when $\tilde{\sigma} = \sigma$. A fundamental property of public key encryption is that it is difficult to generate (i.e., forge) $k_A(A)$ without the key k_A, even given the public key K_A.

---

[9]Public key cryptography is sometimes used to set up even less computationally complex cryptographic frameworks based on symmetric keys or one-way hash functions.

Suppose another actor C has acquired the signature $k_A(A)$ but not the private key k_A, i.e., C has either legitimately received or intercepted prior signed communications from A. To impersonate A, C can mount a *replay* attack by simply appending $k_A(A)$ to a message. Online, there are different approaches that can be used to detect such impersonations. For example, the decrypted identity can be studied in the networking context in which it was received, e.g., it can be checked against the source/return IP address of the signed message if it was transferred using connection-oriented TCP[10]. A Needham-Schroeder cryptographic approach to thwart replay attacks is as follows.

1. A hails B.

2. B randomly generates, stores and sends a session-specific *nonce* R back to A.

3. A sends $k_A(R)$ to B (i.e., A "signs" the nonce).

4. Finally, B decrypts $k_A(R)$ with A's public key and checks against the stored nonce: if they agree, then A's identity has been authenticated.

Other authentication mechanisms employ similar cryptographic techniques, e.g., Kerberos operating in the application layers, Secure Shell (SSH), and Transport Layer Security (TLS). See, e.g., [91] for an introduction to identity authentication and related security issues.

1.5   PEER-TO-PEER (P2P) SOCIAL APPLICATIONS ONLINE

In the balance of this short book, we consider peer-to-peer applications operating online. The "social" element of such applications pertains to the human participation. This can range from inadvertent human action (e.g., the p2p application is a virus which spreads when an end-user unwittingly clicks on a link in a "socially engineered" e-mail or on a Facebook "egg"), to the selection of uplink bandwidth for a p2p file-sharing application such as BitTorrent, and to distributed versions of the types of interactions that now typically occur on centralized social networking sites such as Facebook and MySpace. Recently, peer-to-peer clients for BitTorrent swarm discovery (i.e., "trackerless" BitTorrent) have been proposed based on distributed hash tables (DHTs), *cf.* Ch. 4 and 8.

Contact between end-users over online social networks can be very rapid and inexpensive, unlike social networks mounted on conventional mail or telephony infrastructures (or through physical contact, of course). Dealing with an excessive volume of interactions, less vigilant peers may admit too many social relationships, e.g., Facebook "friends", thereby revealing too much private information to manifest strangers or automata. Greater control over the dissemination of private information may be achieved by a fully decentralized, peer-to-peer framework for social networking. Naturally, decentralized systems still need to enable basic networking functions such as search based on keywords. To this end, some information would need to be shared. In the following, we will explore distributed social systems that are partially to fully decentralized.

---

[10] Otherwise, the UDP messages could simply have forged source IP addresses too. To mitigate forged source IP addresses, some gateway routers engage in *egress filtering* to block outbound packets with source IP addresses that do not belong to their domain.

CHAPTER 2

Graphs

Social ties between humans are complex and multifaceted. A social tie may involve family, profession, political leanings, race and ethnicity, religion, place of origin, education, hobbies, etc. Some social ties are dynamic, slow to develop or fleeting (e.g., accrued trust through prior transactional experiences, aged over time), while others are permanent (e.g., familial ties). So, a social tie between individuals x and y involves a plurality of factors which may be characterizable by a feature vector $w_{x,y}$, where social ties are asymmetrical in general, i.e., $w_{x,y} \neq w_{y,x}$, and $w_{x,y}$ is naturally dependent on corresponding features of the individuals x and y themselves. Metrics between individuals' feature vectors can be formulated and social subgroups (cliques) can then be found, e.g., using hierarchical or K Nearest Neighbors hard clustering or soft (probabilistic) clustering via Expectation-Maximization (EM) on a Gaussian mixture model [51].

A more than one-dimensional feature vector per edge may indicate multiple types of social ties in play within the network; if the different features of a social tie are "independent", then we are basically considering the edge-union of several independent "unidimensional" social graphs, i.e., an edge in the union (multidimensional social tie) graph exists if the corresponding edge exists in at least one unidimensional social graph. We should note that some peers may not allow their activities in one common community of interest (e.g., pornography enthusiasts) to mingle with those of another (e.g., church group), at least not deliberately. Thus, for the purposes of an application, e.g., referrals or search, the composite graph may be more connected than the social network it models.

In this book, we are interested in features that are of the "ratio" type (i.e., neither categorical nor ordinal) so that they can be manipulated as real numbers and compared by difference or quotient. For example, a one-dimensional feature of $w_{x,y} \in [0, 1]$ (or $w_{x,y} \in [0, \infty)$) could be the degree to which x trusts y, where it's well known that trust is, generally, neither symmetrical (so that $w_{x,y} \neq w_{y,x}$) nor transitive. Also, recall in Sec. 1.2 how the "rank" (which we can interpret socially as popularity) of a web site x is deduced from the linkings between them as captured by $P_{x,y}$.

Again, the *underlay* of such a social network could be a centralized server farm, e.g., Facebook or MySpace, facilitating search based on non-standardized keywords. Alternatively, it could be a more fully distributed, peer-to-peer system. Our focus in this book is the latter scenario.

In this chapter, the social network will be represented by a graph where the humans (a.k.a. peers, users, actors) correspond to vertices (a.k.a. nodes) and social ties/associations between them are represented by edges (a.k.a. links) connecting pairs of vertices. An unweighted, bidirectional edge may be present between a pair of vertices in such a graph simply because some type of significant social relationship exists between the users corresponding to the vertices. Alternatively, a graph could have unidirectional edges weighted by a one-dimensional feature vector describing the social tie which the edge represents. Here, an edge weighted zero is equivalent to a non-existing edge. Graphs allow us to model social distance and to consider its effect on, e.g., transactions between peers leveraging referrals based on trust. We cover some basic graph jargon that will be used subsequently and, to simplify matters, we only consider one-dimensional permanent edge weights.

2.1   BASIC GRAPHICAL TERMS

In the following, we will consider a graph $G = (V, E)$ with vertex set V and (unidirectional) edge set E. Qualifications for graphs with unidirectional edges are given in parentheses.

We will denote a (unidirectional) edge $l \in E$ by the (ordered) vertices $(u, v) = l$ it connects with $u, v \in V$ (where in the unidirectional case l is an arrow with u its tail and v is at its head). If $(u, v) \in V$ then we may say that u is directly connected to v or that u and v are neighboring vertices.

A (unidirectional) *path* of *geodesic length* $h \geq 1$ hops connecting two vertices u (origin/source) and v (destination) is a sequence of h edges $\{u_i, v_i\}_{i=1}^{h}$ where $u_1 = u$, $v_h = v$, and $v_i = u_{i+1}$ for all $1 \leq i < h$ (i.e., the head of the ith edge from the source meets tail of the $(i + 1)^{\text{th}}$). If the edges are weighted, then the path length may be defined instead as, e.g., the sum or maximum of its component edge weights.

A graph is *connected* if every (ordered) pair of vertices has a (unidirectional) path connecting them. A graph is *completely* connected if it is connected by paths each consisting of just one edge.

The *shortest distance* between two points of a graph is the length of the shortest path connecting them.

The *size* of a graph G will typically be taken to be $|V|$, the number of vertices.

The *degree* (out-degree or fan-out) of a vertex is the number of its (outbound) edges.

A loop free path is one in which each vertex on the path appears only once. All shortest paths are necessarily loop free.

A vertex of a *tree* is either the root node, an interior node or a leaf node. Each non-root node in the graph is connected to the root by a single loop-free path, so that $|E| = |V| - 1$ (each vertex except the root has an edge between it and its parent). If all edges are bidirectional or undirected, any node can be considered the root.

2.2   SOCIAL GRAPH MODELS AND ATTRIBUTES

In this section, we define and briefly analyze certain graphical attributes, in part because they are considered important social characteristics. This subject matter is covered in far greater depth in, e.g., [23, 49, 53, 109]. Again, our goal is to simply introduce the concepts, so our analysis for a given graph model, regarding the characteristic attributes defined, is far from complete.

DIAMETER VIA TREE APPROXIMATION

The *diameter* of a connected graph can be defined as the *maximum* or *average* among (ordered) pairs of vertices of the length of shortest paths connecting them. The diameter of a fully connected graph is obviously one. Although far from fully connected, the diameter of certain large social graphs of $|V| \gg 1$ nodes has been famously observed to be quite small (on the order of $\log |V|$).

Suppose that the vertices of a graph have average degree $\eta = \sum_{k=1}^{\infty} k p_k$, where p_k is the probability a vertex has degree k. Define a "branching" process $\{X_k, \ k \in \mathbb{Z}^+\}$ on the vertex set V by first selecting one vertex which is generation zero, X_0. Generation k is the set of vertices $X_k \subset V$ which can be reached by one edge (i.e., are neighbors of) a vertex in X_{k-1}. Assume for simplicity that the successive generations are disjoint [115, 117], which may be reasonable up to generation k such that $n_k + \eta \ll |V|$, where $n_k := |X_0 \cup X_1 ... \cup X_k|$. In this case, the number of D-hop neighbors of X_0 in the branching graph can be shown to have mean

$$\tilde{\eta} \eta^{D-1}, \tag{2.1}$$

where $\tilde{\eta}$ is the mean number of neighbors of a somehow typical vertex [117]. One way to compute $\tilde{\eta}$ assumes an *edge* is chosen uniformly at random. Let X_0 be one of its attached vertices[1]. The degree of X_0 is then distributed proportional to $k p_k$, i.e., vertices with higher degrees k are more likely to have one of their edges thus chosen. So, the probability that X_0 has degree k, i.e., that $d_{X_0} = k$, is

$$\frac{1}{\sum_{j=0}^{\infty} j p_j} k p_k \quad = \quad \frac{1}{\eta} k p_k \ =: \ \tilde{p}_k,$$

i.e., $\tilde{\eta} = \sum_k k \tilde{p}_k$. Note that the mean number of vertices (2.1) thus discovered forms a tree of diameter $2D$. So under the above assumptions, if $|V| \approx \tilde{\eta} \eta^{D-1}$ then

$$2D \quad \approx \quad 2 + 2 \log_\eta (|V|/\tilde{\eta}).$$

The probability of disjointness of successive generations, however, diminishes as $n_k + \eta \to |V|$, owing to the increasing probability of revisiting a previously discovered vertex by the branching process or discovering the same vertex multiple times in one generation, see [53] (p. 71 at bottom) and [170].

The number of vertices in the tree is an upper bound on any graph consisting of the same number of nodes possessing such a degree distribution. Although tree-based calculations may only be approximate, the conclusion that the diameter is $O(\log |V|)$ can be proved in many cases, see e.g., Ch. 2 of [53] for the classical Erdos-Renyi random graph (with the parameters $|V|$ and η, each of the $\binom{|V|}{2}$ possible bidirectional edges is independently present with probability $\eta/(|V| - 1)$). In the following, we will derive similar results for a "small world" graph and for "structured" graphs, also see Sec .4.2.

---

[1]So, we choose to bias our measurement of diameter to path-lengths centered on more popular vertices.

POWER-LAW DEGREE DISTRIBUTION OF PREFERENTIAL ATTACHMENT GRAPHS

When a graph is said to follow a "power law", the tail of its degree distribution decays only polynomially quickly, i.e.,

$$P(d = x) \quad \propto \quad x^{-\rho}$$

for some $\rho > 0$ and for all sufficiently large integers x. Such a degree distribution (i.e., with specific values of ρ) has been empirically observed in many networking settings [25, 49, 53].

In [25, 139], the following preferential-attachment iteratively generative graph model was described employing a single strictly positive parameter $\alpha < 1$. The graph grows in such a way that at each iteration:

- with probability $1 - \alpha$ a unidirectional edge is generated at each iteration connecting a pair of existing, randomly selected vertices, where the probability that the new edge is outbound from a vertex of out-degree i is proportional to the current number of such vertices;

- otherwise (i.e., with probability α), a vertex with a single outbound edge is generated, i.e., the new vertex has out-degree one and is connected to an existing vertex selected at random.

We set the initial graph $G_1 = (V_1, E_1)$, i.e., after iteration $k = 1$, to have one vertex and one self-edge.

As a consequence of the following result, the out-degree distribution of $G_k = (V_k, E_k)$ approaches a power law with parameter $\rho = 1/(1 - \alpha)$ as $k \to \infty$.

Claim 2.1   The out-degrees of graph G_k converge in distribution to beta with parameter $\rho := 1/(1 - \alpha)$.

Proof. After iteration $k \geq 1$, let $F(i, k)$ be the expected number of vertices with out-degree $i \in \{1, ..., k\}$ of the graph $G_k = (V_k, E_k)$. Thus,

$$\sum_{i=1}^{|E_k|} i F(i, k) \quad = \quad |E_k|, \quad \text{where } |E_k| \equiv k. \tag{2.2}$$

By first conditioning on G_{k-1} and then taking expectations, we obtain the following iterative dynamics for F:

$$F(i, k + 1) - F(i, k) \quad = \quad \frac{(i - 1)F(i - 1, k) - i F(i, k)}{Z_k} \quad \text{for } i \geq 2 \tag{2.3}$$

and

$$F(1, k + 1) - F(1, k) \quad = \quad \alpha - \frac{F(1, k)}{Z_k}, \tag{2.4}$$

where Z_k is defined so that $i F(i, k)/Z_k$ is the probability that the edge chosen in iteration k is outbound from an *existing* vertex that has out-degree $i \geq 1$. So,

$$1 - \alpha = \sum_{i=1}^{k} i F(i, k)/Z_k$$
$$\Rightarrow Z_k = \frac{k}{1 - \alpha} \text{ by (2.2).}$$

Note that in (2.3) and (2.4), an edge "arrives" at iteration k so that with probability $i F(i, k)/Z_k$ the population of vertices of out-degree i decreases by one, and thus the population of vertices with out-degree $i + 1$ increases by one.

As $k \to \infty$, assume the "equilibrium condition" that

$$b(i) := F(i, k)/F(i - 1, k) \text{ is independent of } k.$$

Substituting this condition into (2.3) and (2.4) and solving gives

$$b(i) = \frac{i - 1}{i + \rho} = \frac{F^*(i)}{F^*(i - 1)},$$

where $F^*(i) := \lim_{k \to \infty} F(i, k)$ for all integers $i \geq 1$. So, for all $i \geq 2$,

$$\begin{aligned}
F^*(i) &= b(i)b(i - 1)...b(2)F^*(1) \\
&= \frac{(i - 1)!}{(i + \rho)(i - 1 + \rho)...(1 + \rho)} F^*(1) \\
&= \frac{\Gamma(i)\Gamma(1 + \rho)}{\Gamma(i + 1 + \rho)} F^*(1) \\
&= B(i, \rho + 1)F^*(1),
\end{aligned}$$

where Γ is the gamma function and B is the beta function.

□

Since $\Gamma(i)/\Gamma(i + \rho + 1) \sim i^{-\rho}$ for large i, the beta distribution is a power law. Additional details regarding the model (2.3) and (2.4) are given in [139]. Note that if ρ is a positive integer (recall $\alpha = 1 - \frac{1}{\rho}$ by definition), $\Gamma(i + 1 + \rho) = (i + \rho)!$ and $B(i, \rho + 1) = \binom{i+\rho}{i-1}^{-1}$.

"Popular" vertices[2] form a densely connected subnetwork. This naturally leads to a low diameter of the largest connected component and increases the likelihood that the overall graph is connected, see below.

CLUSTERING COEFFICIENT AND SMALL WORLDS GRAPHS

Consider a graph of $n := |V|$ vertices with bidirectional edges. Suppose i belongs to a *neighborhood* of $n_i \leq n$ vertices. The neighborhood has $\binom{n_i}{2} := n_i(n_i - 1)/2$ possible internal edges. Let d_i be

---

[2]Such as hub vertices with high degree which will tend to arrive at earlier iterations in the preferential attachment model.

the actual number of edges between nodes of this neighborhood. The local clustering coefficient of i's neighborhood can be defined to be

$$C_i \ := \ d_i \Big/ \binom{n_i}{2}. \tag{2.5}$$

For a fully connected graph with $n_i = n$ for all vertices i, clearly $C_i = 1$ for all i. The clustering coefficient of the whole graph can be defined as the average of the C_i over all of its vertices i.

Again, in the classical Erdos-Renyi [55] random graph with n vertices, each of the $\binom{n}{2}$ possible (bidirectional) edges are independently present with probability $p(n) \sim \eta/(n-1)$ for a constant $\eta > 0$. If a vertex i shares an edge with j and another with l, then it is not very likely in a large Erdos-Renyi graph (only with probability $p(n)$) that j and l also share an edge. Such simple "triangle clusters" are more likely to exist in social networks (this said, certain aspects of social association, e.g., degree of trust or influence, are generally neither symmetric nor transitive).

The Watts-Strogatz *small worlds* random graph model [152] has larger clustering coefficient than an Erdos-Renyi graph of the same diameter. To construct this graph, first suppose that the vertices can be mapped (embedded) onto a unit-circumference circle, and spaced apart equally, in such a way that all vertices in any arc of certain length α (i.e., local, neighboring vertices) share a bidirectional edge. Thus, the degree of each vertex is about αn and the total number of edges is $\alpha n^2/2$, where $\alpha n \geq 4$ is required. To obtain the Watts-Strogatz random graph, edges are now independently "rewired" with probability r, i.e., new vertices are chosen for the edge uniformly at random from the $\binom{n}{2}$ vertex pair population. This rewiring creates "long distance" neighbors in the graph.

For rewiring probability $r = 0$, the graph clearly has diameter about

$$\frac{1/2}{\alpha/2} \ = \ \frac{1}{\alpha}.$$

For large r, the graph begins to resemble that of Erdos-Renyi with edge probability $p = \alpha n$. One can compute the normalized diameter $D(r)/D(0)$ (mean minimum geodesic path lengths among pairs of vertices) and the mean clustering coefficient, $C(r)/C(0)$, as functions of r. As r increases from zero to one, the $D(r)/D(0)$ decreases (due to the short-cuts afforded by rewiring to long-distance neighbors) much more rapidly than $C(r)/C(0)$, giving an interval of values of r where diameter is low and clustering coefficient is high [49].

In Sec .4.2 we will show that the diameter of such a random graph with "structure" has small diameter by a greedy routing/search algorithm that actually finds short paths. See [45] for the clustering properties of four example online graphical datasets.

COMPONENTS (SUBGRAPHS) THAT ARE CONNECTED

A principal result for the Erdos-Renyi random graph [55] states that if $\eta < 1$ then the number of vertices in the largest connected component is $S(n) = O(\log n)$ almost surely (a.s.)[3]. Alternatively,

---

[3]i.e., with probability one as $n \to \infty$.

if $\eta = 1$ then $S(n) = O(n^{2/3})$ a.s. or if $\eta > 1$ then $S(n) = O(n)$ a.s., where in the latter case the component has a large fraction of the total number of vertices in the graph and hence is "giant" sized. The distributions of the number of connected components and their sizes have been studied for other types of random graph models, including those that have significantly higher clustering coefficients than Erdos-Renyi graphs, see e.g., [53]. Similar "criticality" results have be obtained for random percolations on graphs, see, e.g., the introduction in Ch. 3 of [66].

To study the connectivity for a graph with "location-based" clustering, suppose the vertices of a social graph can be somehow mapped to points in a k-dimensional hypercube $[0, 1]^k$ in a way that if a pair of vertices x, y are sufficiently close, say by Euclidean distance $| \cdot |$, then a bidirectional edge exists between them. An interpretation of this "embedding" of a social graph into $[0, 1]^k$ is that the k-vector $x \in [0, 1]^k$ represents k real-valued social attributes of the corresponding user. One such attribute may explicitly indicate status in a particular society. If categorical or ordinal features are also involved, a more general asymmetrical "dissimilarity" metric $D(x, y)$ could instead be used to indicate the tendency of two peers to be linked together in a social relationship.

Claim 2.2   For a graph $G_{n,r}$ of n vertices independently and uniformly located in $[0, 1]^k$ at random, with a bidirectional edge between vertex x and y if and only if $|x - y| < r$ for some $r < 1$: if

$$r \geq \left(c\frac{\log n}{n} \right)^{1/k} \tag{2.6}$$

for a constant $c > k^k$, then

$$\lim_{n \to \infty} P(G_{n,r} \text{ is connected}) = 1.$$

Proof. Partition the hypercube into smaller ones (bins) of size $(r/k) \times (r/k) \times \ldots \times (r/k)$. Thus, there are $(k/r)^k$ such bins. The probability that a bin is empty is $(1 - (r/k)^k)^n \leq \exp(-n(r/k)^k)$, where the inequality is $1 - x \leq e^{-x}$ for $x > 0$. So, the number of empty bins X has mean

$$\begin{aligned} EX &= (k/r)^k (1 - (r/k)^k)^n \\ &\leq (k/r)^k \exp(-n(r/k)^k). \end{aligned}$$

Graph $G_{n,r}$ is connected if no bin is empty, so $P(G_{n,r} \text{ is connected}) \geq P(X = 0)$. Thus,

$$\begin{aligned} P(G_{n,r} \text{ is not connected}) &\leq 1 - P(X = 0) \\ &= P(X > 0) \\ &\leq EX \\ &\leq (k/r)^k \exp(-n(r/k)^k), \end{aligned}$$

where the second inequality is Markov's:

$$EX = \sum_{i=1}^{(k/r)^k} i P(X = i) \geq \sum_{i=1}^{(k/r)^k} P(X = i) = P(X \geq 1).$$

Finally, use (2.6) to show $P(G_{n,r}$ is not connected$) \to 0$ as $n \to 1$. $\square$

SOCIAL IMPORTANCE BASED ON PATHS

There are many measures of social importance of a vertex i, e.g., the local clustering coefficient (2.5) or the globally normalized degree d_i/η, again where η is the mean vertex degree of the entire graph (obviously, the variance of the degree or the maximum degree could be considered as well). Other measures are based on paths in the social graph. Let σ_{jk} be the number of different *shortest* paths between vertices $j \neq k$. For $j \neq i \neq k$, let $\sigma_{jk}(i)$ be the number of paths in σ_{jk} in which the node i resides, i.e., i is between j and k on a shortest path. The *betweenness centrality* [59] of node i is

$$\sum_{j,k:\ j\neq i\neq k} \frac{\sigma_{jk}(i)}{\sigma_{jk}}. \tag{2.7}$$

Such path-based measures are relevant to, e.g., social reputation systems which are based on referral chains/paths, as discussed subsequently in this book. Generalizations of these quantities to weighted, unidirectional edges are straightforward.

TIES BETWEEN SOCIAL GROUPS

We can also assess the importance of social *groups* in similar ways. Consider two *disjoint* groups $A, B \subset V$ and define the normalized strength of the collective ties (or "correlation") between them as

$$W_{A,B} = \frac{\sum_{i \in A,\ j \in B} w_{i,j} w_{j,i}}{|A| \cdot |B|},$$

where $w_{i,j}$ is the weight of a unidirectional edge (i, j), here simply indicating whether an edge exists from vertices i to j, i.e., $w_{i,j} \in \{0, 1\}$. Recall that d_i is the edge-degree of vertex (peer) i, i.e., $d_i := \sum_{j \neq i} w_{i,j}$. For any (social group) $G \subset V$, define the "degree weighted homophily" [64]:

$$H(G) := \frac{W_{G,G} + W_{G^c,G^c} - 2W_{G,G^c}}{|G|^{-2} \sum_{i \in G} d_i^{-1} + |G^c|^{-2} \sum_{i \in G^c} d_i^{-1}}, \tag{2.8}$$

where G^c is the complement of G so that $|G^c| = n - |G|$ when $n := |V|$. Note that H increases with the number of connections (edges) between peers who are both in the same group (G or G^c), and decreases with connections from a peer in one group to a peer in the other. For the special case of a regular graph (all vertices with equal degree so that $d_i = d$ for all i), $H(G)$ is just the ratio of the difference between the total number of intra-group edges and inter-group edges to the total number of edges in the graph, the latter being $nd/2 = \sum_i d_i$.

Again, these quantities can be generalized to graphs with unidirectional edges with asymmetric non-Boolean weights. These and other graphical characteristics can be related to the performance of certain processes operating on social networks, such as single-threaded search, the propagation of beliefs (consensus), or epidemics, e.g., [45, 97]. Finally, we mention that it's possible that social groups are not disjoint [116, 118, 151], but, again, this does not necessarily imply that groups are functionally correlated through their common peers. We could model a peer with, say, two vertices, one in each of the two social groups to which she belongs, and connect these two vertices by an edge only if this peer acts as a bridge between them.

2.3 DISCUSSION: EDGE AND VERTEX CHURN

Online social networks are dynamic due to the evolving nature of social ties and individuals joining (respectively leaving) the network as they log in (respectively log out or are disconnected due to some kind of time-out or network failure). Such topological churn is more problematic for the basic functioning of a distributed peer-to-peer social network than for one mounted on a single site, e.g., Facebook or MySpace, and so these networks need to be designed robustly with churn in mind.

There are several different models of peer churn, e.g., a system with fixed online population of peers with peer "reset" corresponding to one peer's departure *coupled* to a new peer's arrival [57]. Also, see models based on spatial stochastic point processes [15].

Consider a social network with a small but critical mass of hub peers (also known as super-peers). Most shortest paths in the graph pass through one or more hubs, i.e., the hub peers i have very high betweenness centrality (2.7). Because of the presence of super-peers, such graphs have heavy-tailed degree distributions. Not surprisingly, connectivity and diameter of such graphs are robust in the face of removal of randomly selected edges or non-hub peers (as compared to, e.g., Erdos-Renyi graphs), but perhaps more sensitive to the removal of hub peers [24].

Churn-related threats to large-scale peer-to-peer systems include deliberately causing churn to magnify the overhead required to deal with it (i.e., join-leave attacks) and peer departures due to the spread of epidemics, e.g., [170]. Clearly, epidemics will spread more rapidly if hub peers are infected and if the subset of vulnerable nodes are well-connected, *cf.* Ch. 11.

CHAPTER 3

Games

In games, individual player's (user's) behavior will be modeled by a payoff (utility) function. Utility minus the cost of playing the game will be called net utility. Each player is assumed to have an action/play space of decisions. These actions are coupled through network dynamics. The utility of a given user will depend, at least in part, on all other user actions. Our games will involve iterative plays. A player's objective will be to maximize their net utility.

Game-theoretic models for telecommunication systems have recently been surveyed in [7]. In this chapter, we review simple non-cooperative games among greedy players (selfishly focusing on their own utility), followed by a discussion of games with collaborative behavior. Generally, more complex engagement models than those illustrated below have been considered, including asynchronous players with limits on, or costs associated with, player observation and communication (to mount collaborative action).

3.1 SET-UP FOR NON-COOPERATIVE GAMES

We have already considered a simple game involving greedy and non-cooperative players in the iterative TCP flow control example of Sec .1.1. In our framework for such games, the ith user's play, q_i, is focused on maximizing their own net utility,

$$U_i(\gamma_i(\underline{q})) - M\gamma_i(\underline{q}),$$

where:

- $\underline{q}$ is the vector of all current plays of the different users involved in the game;

- $\gamma_i(\underline{q})$ represents what the network provides user i at $\underline{q}$;

- U_i is user i's utility, i.e., how user i values $\gamma_i(\underline{q})$; and

- here we take the *usage-based* cost of $\underline{q}$ as $M\gamma_i(\underline{q})$ using the network-set price M.

NASH EQUILIBRIUM

A principle objective of these games is to find a stalemate situation, called a Nash equilibrium: a point (collective play) $\underline{q}^*$ where any unilateral deviation q_i from q_i^* (i.e., the collective play $(q_i; \underline{q}_{-i}^*)$ representing $\underline{q}^*$ with q_i instead of q_i^*) will result in a lower net utility for player i than that at $\underline{q}^*$. That is, for all i, q_i^* is a maximum of

$$U_i(\gamma_i(q_i; \underline{q}_{-i}^*)) - M\gamma_i(q_i; \underline{q}_{-i}^*)$$

viewed as a function of q_i. If the U_i are strictly concave, i.e.,

$$U_i''(\gamma) \quad < \quad 0 \text{ for all } \gamma, i, \tag{3.1}$$

then for the particular type of pricing applied here we see that the optimal/desired

$$\gamma_i^* \quad = \quad (U_i')^{-1}(M) \quad =: \quad y_i.$$

In the following, we typically assume

$$\frac{\partial \gamma_i}{\partial q_i}(\underline{q}) \quad > \quad 0 \text{ for all } \underline{q}, i, \tag{3.2}$$

$$\frac{\partial \gamma_i}{\partial q_j}(\underline{q}) \quad < \quad 0 \text{ for all } \underline{q}, i \neq j. \tag{3.3}$$

So, we can define

$$F_i(\underline{q}) \quad := \quad \arg\max_q U_i(\gamma(q; \underline{q}_{-i})) - M\gamma(q; \underline{q}_{-i}). \tag{3.4}$$

For example, for the ALOHA medium access game of [73, 74], player i's (re)transmission probability is q_i so that the stationary throughput of user i

$$\gamma_i(\underline{q}) \quad = \quad q_i \prod_{j \neq i}(1 - q_j) \text{ and thus}$$

$$F_i(\underline{q}) \quad = \quad \frac{y_i}{\prod_{j \neq i}(1 - q_j)}.$$

The result is a *quasi-stationary* game, i.e., player actions are based on observed performance in steady-state equilibrium[1].

For a more abstract example relevant to the subsequent discussion in this book, consider a group of players interacting online. The outcome of a transaction between a pair of players results in an adjustment of the "reputation" of one from the point of view of the other. These "direct" reputations can be can be localized[2] or they can be shared among users to form the basis of referrals, *cf.* Ch. 6. Direct reputations or indirect referrals can affect whether requested transactions between a particular pair of peers will occur in the future. In this way, reputation/referrals track past behavior and have consequences in the future in terms of rewards and penalties.

JACOBI ITERATION FOR CONTINUOUS ACTION SPACE

The following continuous-time Jacobi iteration approximates the above "best-response" discretely-iterative dynamics with "better-response" [60] continuous dynamics:

$$\dot{\underline{q}}(t) \quad = \quad \underline{F}(\underline{q}(t)) - \underline{q}(t), \tag{3.5}$$

---

[1] Also known as "fictitious play" [28], this assumption is common to "evolutionary" game frameworks [154].
[2] As in BitTorrent file sharing for purposes of intra-swarm choking rules, *cf.* Ch. 8.

where user actions are assumed to be synchronized. So by (3.4), a stationary or equilibrium point of (3.5) (a point where $\dot{q} = 0$) will be a Nash equilibrium point. The rate can be changed in the modified dynamics $\dot{q}(t) = \kappa[\underline{F}(q(t)) - q(t)]$ by varying the constant $\kappa > 0$ (or a positive-definite diagonal matrix κ if action-rate/responsiveness varies among the players).

In a distributed system, players may take small steps toward their currently optimal play (based on their current knowledge of the state of the game, in particular the actions of the other players). One reason for this is that players act simultaneously so the optimal plays may change significantly. Thus, small steps may avoid large oscillations in the network dynamics, i.e., oscillations that will be harmful to performance (recall the balance between tentative and aggressive action taken by TCP in its distributed congestion control strategy). Also, small steps may ensure convergence to stable interior equilibria, even under asynchronous updates by players operating at different action rates [17, 19], and avoid deadlock boundary equilibria. Other types of continuous-time models for competitive games exist, e.g., replicator dynamics for games among population aggregates [154].

Often the play-action space is limited, e.g., $q_i \in [0, q^{\max}]$ for all i so that (3.5) holds only in the interior of the hypercube, $q \in [0, q^{\max}]^n$, where n is the number of players involved. Again, Nash equilibria can exist on the boundaries of play-action space. A Nash equilibrium point (NEP) q^* is locally asymptotically stable if there exists a neighborhood[3] about q^* such that beginning at *any* $q(0)$ in the neighborhood (3.5) will converge to q^*. When multiple Nash equilibria exist, each may have its own type of local stability, e.g., [3, 73, 74, 123, 138, 164]. The outcomes of games are typically assessed at stable Nash equilibria.

One can apply the Hartman-Grobman theorem to test for the stability of (3.5) at a given interior NEP q^*: q^* is locally asymptotically stable if all eigenvalues of the Jacobian $\frac{\partial \underline{F}(q) - q}{\partial q}(q^*) = \frac{\partial \underline{F}}{\partial q}(q^*) - \mathbf{I}$ have negative real part.

There may exist a local Lyapunov function for the system (3.5), i.e., a function Λ such that

$$\frac{d\Lambda}{dt}(q) = \langle \nabla \Lambda(q), \dot{q} \rangle = \langle \nabla \Lambda(q), \underline{F}(q) - q \rangle \le 0, \quad \text{for all } q$$

where $\langle \underline{u}, \underline{v} \rangle := \sum_i u_i v_i$ is the inner (or dot) product of the vectors $\underline{u}$ and $\underline{v}$ (so that the equality in the display above is just the chain rule). Thus, (3.5) will evolve so as to converge to local minima of Λ. A NEP q^* is asymptotically stable if all eigenvalues of the Hessian of Λ at q^* have positive real part, i.e., Λ is locally convex at q^*.

An equilibrium point q^* is *Pareto* dominant if *each* player's net utility at q^* is not less than that at any other equilibrium. Nash equilibria are not necessarily Pareto.

3.2   SYMMETRIC ALTRUISM

In this section, we assume players are aware of one another's net utilities and are partially inclined to cooperate [83]. As is common in some wireless settings, e.g., tactical mobile ad hoc networks

---

[3]An open ball in $[0, q^{\max}]^n$ of positive radius.

(MANETs), network nodes may engage in fully or partially cooperative behavior with respect to some of their peers. Altruistic action can be for the purpose of routing [14], medium access, etc. In this section, we consider a symmetric situation where the players have similar

- priorities of communications and

- degree to which their behavior is altruistic.

This said, the *demands* of the players are, however, generally assumed different (asymmetric) in the following.

   In distributed systems, altruistic actions can easily be shown to be counter-productive in the presence of :

- limited information and/or observation errors;

- excessive communication overhead to (securely) convey more accurate network state information; and/or

- deliberate injection of false observation information by enemy actors.

In these cases, selfish behavior may give better performance. In coalitional games, the cost of co-operation is typically weighed against its benefits, e.g., [10]. It is possible, however, that symmetric altruistic behavior can be beneficial to stabilize feasible equilibria. In the following, we give an example for distributed power-control medium-access dynamics having a single feasible equilibrium.

ALTRUISTIC GAME SET-UP

Here, by using his/her action q_i, suppose the ith user seeks to maximize the composite net utility

$$\sum_j \alpha_{ij}(U_j(\gamma_j) - M\gamma_j), \qquad (3.6)$$

where i and j index the players and we assume:

- (3.1)-(3.3), and

- all $\underline{\alpha}_i$ are in the standard simplex, i.e.,

$$\sum_j \alpha_{ij} = 1 \quad \text{with} \quad \alpha_{ij} \geq 0 \quad \text{for all } i, j. \qquad (3.7)$$

   So, the users need to be made aware of each others' actions and utilities[4].

---

[4]In the purely selfish/non-cooperative games ($\alpha_{ii} = 1$ for all i) of [75], actions were taken based on observations in quasi steady-state thus not requiring any coordination between presumed selfish/non-cooperative peers.

Purely selfish/non-cooperative games

The first derivative of the net utility of player i with respect to player i's control parameter q_i is

$$(U_i'(\gamma_i(\underline{q})) - M)\frac{\partial \gamma_i}{\partial q_i}(\underline{q}),$$

where we have now explicitly written the γ_i as functions of $\underline{q}$ and taken $\alpha_{ii} = 1$ for all i. So, at a Nash equilibrium point (NEP) $\hat{\underline{q}}$ of a *non*-altruistic game, i.e., where $\alpha_{ii} = 1$ for all i,

$$\gamma_i(\hat{\underline{q}}) \;=\; (U_i')^{-1}(M) \;=:\; y_i \quad \text{for all } i. \tag{3.8}$$

Note that by the concavity assumption of the U_i, the second derivative of the net utility of player i *at* a NEP $\hat{\underline{q}}$ is

$$U_i''(\gamma_i(\hat{\underline{q}}))\left(\frac{\partial \gamma_i}{\partial q_i}(\hat{\underline{q}})\right)^2 \;<\; 0.$$

That is, (3.8) and (3.1) are conditions for $\underline{q}$ to be a NEP.

Altruistic invariance of interior NEPs

Claim 3.1   Under (3.1), the NEPs of the purely selfish game ($\alpha_{ii} = 1$ for all i) are also NEPs under the altruistic objective (3.6) for all $[\alpha_{ij}]$ satisfying (3.7).

Proof. Note that for objectives (3.6), the first-order optimality condition for player i is

$$\sum_j \alpha_{ij}(U_j'(\gamma_j(\underline{q})) - M)\frac{\partial \gamma_j}{\partial q_i}(\underline{q}) \;=\; 0.$$

So, clearly this condition is satisfied when $\underline{q} = \hat{\underline{q}}$, i.e., under (3.8). Moreover, the second-order condition at such a $\hat{\underline{q}}$ (again, under (3.8)) reduces to

$$\sum_j \alpha_{ij}U_j''(\gamma_j(\hat{\underline{q}}))\left(\frac{\partial \gamma_j}{\partial q_j}(\hat{\underline{q}})\right)^2 \;<\; 0.$$

$\square$

   In the following example, we will consider "symmetric" altruism parameters α, i.e., with α_{ii} constant over i and $\alpha_{ji} = \alpha_{ij}$ for all $i \neq j$, as for the case of equal-priority communication.

   Again, it is possible that NEPs are not asymptotically stable. Although the positions of the NEPs may not change under altruistic behavior, the nature of their stability may change with the (α_{ij}) parameters.

WIRELESS MEDIUM ACCESS BY POWER CONTROL

Game-theoretic models for wireless medium access by power control have been extensively studied, e.g., [6, 107, 135, 136, 158], including consideration of issues of robust convergence to equilibria, e.g., [5, 106]. In the following, we consider a game played by unidirectional transmitted data *flows* between pairs of nodes. We assume that a transmission attempt occurs in every time-slot by every player.

The signal to interference and noise ratio (SINR) of the ith flow is

$$\text{SINR}_i(\underline{q}) \quad := \quad \frac{q_i h_{ii}}{N + \sum_{j \neq i} h_{ji} q_j},$$

where N is the ambient noise power, the power $q_i \geq 0$ pertains to the transmitter of flow i, and h_{ji} are the path gains between the transmitter of flow j and the receiver of flow i. So, SINR_i is the SINR at the receiver of flow i. See Figure 3.1 illustrating two flows with transmitters T_k and receivers R_k, $k \in \{0, 1\}$. The demands of each player i are $y_i := (U_i')^{-1}(M)$ as above.

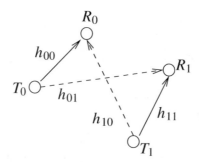

Figure 3.1: Two interfering flows.

Note that bidirectional links are not considered above due to self-interference at the transmitters/receivers. Typically, each-way communication of a bidirectional link will be separated by TDMA, FDMA, or CDMA/spread-spectrum means. In the TDMA setting for a distributed multihop wireless network, a spatial scheduling problem ensues, e.g., [114, 132].

From SINR to QoS γ

For different technologies, we can idealize

$$\gamma_i(\underline{q}) \quad := \quad \Gamma(\text{SINR}_i(\underline{q}))$$

for correspondingly different increasing functions Γ, so that (3.2) holds. Assuming that the bit error events are independent, the probability of correct reception of an n-bit packet is

$$\Gamma(\text{SINR}) = (1 - p_e(\text{SINR}))^n,$$

where p_e is bit error probability. For example, if n is large and p_e has a decaying exponential form (e.g., when using DBPSK or GFSK modulation [126]), then we can approximate

$$\Gamma(\mathsf{SINR}) \approx \exp(-n\exp(-\mathsf{SINR})).$$

Alternatively, Shannon's expression for capacity $\log(1 + \mathsf{SINR})$ is often used to map SINR to service quality.

The selfish game

In a quasi-stationary selfish game, the users observe their interference and the ith user acts according to

$$q_i \;=\; \Upsilon_i \cdot (N + \sum_{j \neq i} q_j h_{ji}) \;=:\; F_i(\underline{q}),$$

where

$$\Upsilon_i \;:=\; \Gamma^{-1}(y_i)/h_{ii}. \tag{3.9}$$

Note that this system is simply affine in $\underline{q}$ and a unique NEP $\underline{q}$ such that $\underline{q} = \underline{F}(\underline{q})$ can be determined[5] if the matrix $\mathbf{I} - \Psi$ is non-singular where $\Psi_{ji} := h_{ji}\Upsilon_i$ for all $j \neq i$ and $\Psi_{ii} := 0$ for all i. That is, the NEP would be

$$\underline{q} \;=\; (\mathbf{I} - \Psi)^{-1} N\underline{\Upsilon}. \tag{3.10}$$

For a two-player game indexed $i \in \{1, 2\}$, a Lyapunov function of (3.5) is the quadratic form

$$\Lambda(\underline{q}) \;=\; \sum_i h_{i,3-i}\Upsilon_{3-i}(\frac{1}{2}q_i^2 - N\Upsilon_i q_i) - \prod_i q_i h_{i,3-i}\Upsilon_i.$$

The NEP is stable if

$$\prod_i h_{i,3-i}\Upsilon_i \;<\; 1. \tag{3.11}$$

The purely altruistic game

For an altruistic two-player game with information sharing as above, the ith user acts according to

$$q_i \;=\; \frac{1}{h_{3-i,i}}(\frac{q_{3-i}}{\Upsilon_{3-i}} - N) \;=:\; \tilde{F}_i.$$

---

[5]Due to power and fairness constraints, a centralized base station might need to find a fair "minimum norm" set of powers in a situation where no feasible Nash equilibrium solution exists.

Note that $\tilde{F}_i$ is also a simple affine function. Again, we can show that the NEP (3.10) holds here too and similarly study its stability properties as in the non-cooperative case. The Lyapunov function for the purely altruistic ($\dot{\underline{q}} = \tilde{\underline{F}} - \underline{q}$) two-player game is

$$\tilde{\Lambda}(\underline{q}) \;\; = \;\; \sum_i \frac{1}{h_{i,3-i}\Upsilon_i}\left(\frac{N}{h_{3-i,i}}q_i + \frac{1}{2}q_i^2\right) - \prod_i \frac{q_i}{h_{i,3-i}\Upsilon_i},$$

and the NEP is stable if

$$\prod_i h_{i,3-i}\Upsilon_i \;\; > \;\; 1, \tag{3.12}$$

which obviously cannot be true if (3.11) holds.

Recalling (3.9), we see a *potentially* beneficial role for symmetric altruism in the case where $\prod_i \Gamma^{-1}(y_i) > \prod_i h_{ii}/h_{i,3-i}$ so that (3.11) does not hold, and therefore (3.12) does.

Numerical example
Consider an example where the frame sizes $n = 1024$ bits (128 bytes), the desired mean correct frame transmission probabilities $(y_0, y_1) = (.97, .98)$, the noise power $N = 1.0$ (so the transmission powers (q_0, q_1) are normalized with respect to noise), the path gains[6] $h_{i,i} = 0.1$ and $h_{i,3-i} = 0.005$ for all i, and the utilities are of the arctan form,

$$U_i(\gamma) \;\; = \;\; \frac{2}{\pi}U_i^{\max}\arctan(\beta_i\gamma)$$

for positive constants $U_i^{\max}$, β_i.

The contours of the Lyapunov functions Λ and $\tilde{\Lambda}$ for this idealized scenario are respectively depicted in Figures 3.2(b) and 3.2(a). The unique feasible NEP is $(q_0, q_1) = (224, 230)$ with corresponding SINRs of 10.4 and 10.8, respectively, and comparable noise and interference magnitudes. This NEP does not change position as the altruism parameter α changes (Claim 3.1). The asymptotic stability condition for altruistic dynamics (3.12) does *not* hold for this example, giving the saddle contour for $\tilde{\Lambda}$ in Figure 3.2(a). Stability therefore does hold for selfish dynamics (3.11), as can be seen in Figure 3.2(b).

More complex channel/engagement models, with limits observation and actuation (play-space), have been considered in this context, see, e.g., [106].

3.3   GAMES WITH COLLECTIVE OUTCOMES

In the games described above, outcomes of individual players were characterized by utilities and this lead to the notion of a Nash equilibrium. We now consider outcomes that are collective among the players.

---

[6]The coding strategy will, in many cases, additionally reduce the interference factor $h_{i,3-i}$ beyond propagation attenuation, i.e., by a "processing (power) gain" factor of the code, here assumed to be 13dB $= 20 = h_{i,i}/h_{i,3-i}$.

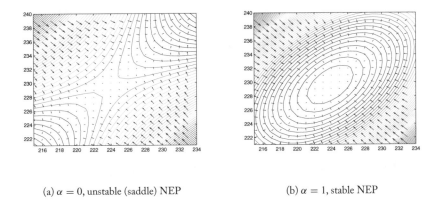

(a) $\alpha = 0$, unstable (saddle) NEP   (b) $\alpha = 1$, stable NEP

Figure 3.2: Power control dynamics.

A SIMPLE TWO-PLAYER EXAMPLE

Suppose a customer group consumes a service which needs to be mounted by two cooperating vendors/players, $i \in \{1, 2\}$. For example, player 1 could be an Internet service provider (ISP) and player 2 could be a content provider. For simplicity, assume a common customer demand profile d that diminishes linearly with price,

$$d(\underline{q}) \;=\; d_0 - c(q_1 + q_2),$$

where q_i is the price leveed by player i and d_0 is the maximum demand, e.g., [29, 54]. So, the payoff of player i is quadratic $U_i = q_i d(\underline{q})$.

If the players do not cooperate, the Nash equilibrium is

$$\underline{q}^* \;=\; (\frac{d_0}{3c}, \frac{d_0}{3c}).$$

The individual utilities at this point are $U_i = d_0^2/(9c)$, so the maximal collective utility is $2d_0^2/(9c)$.

If the players cooperate and coordinate, then to maximize the collective utility $U_1 + U_2 = d_0 - cq$ over $q := q_1 + q_2, q^* = d_0/(2c)$ is chosen. The maximal collective utility is now $d_0^2/(4c) > 2d_0^2/(9c)$, i.e., larger than in the non-cooperative case. This reduction through non-cooperation of the collective utility is a simple example of a "price of anarchy" [131].

DIVIDING COLLECTIVE OUTCOMES BY SHAPLEY VALUES

Given a collective outcome, we can try to reckon how to "fairly" divide or attribute the gain U among the participating players. Here, let $U(S)$ be the communal payoff when the subset S of possible players is participating in the cooperative game. Suppose U is a non-negative super-additive function on the set of subsets (i.e., the power set) of $n = |\mathcal{N}|$ players indexed $\mathcal{N} = \{1, 2, ..., n\}$, i.e., $U(\varnothing) = 0$

and for all disjoint $S_1, S_2 \in \mathcal{N}$, $U(S_1 \cup S_1) \geq U(S_1) + U(S_2)$. The *Shapley values* are an n-vector $\underline{v}$ with i^{th} component

$$v_i = \sum_{s=0}^{n-1} p(s,n) \sum_{S \subset \mathcal{N}:|S|=s;\ i \notin S} [U(S \cup \{i\}) - U(S)] \tag{3.13}$$

where

$$p(s,n) = \frac{s!(n-s-1)!}{n!}. \tag{3.14}$$

By considering the payoffs from all possible combinations of players, the Shapley values are a way to partition $U(\mathcal{N})$, so that player i receives the share v_i: i.e.,

$$U(\mathcal{N}) = \sum_{i=1}^{n} v_i. \tag{3.15}$$

To see that (3.15) holds, note that in $\sum_{i=1}^{n} v_i$ the coefficient of $U(S)$ with $s = |S|$ is $sp(s-1,n) - (n-s)p(s,n)$. Since we assume general super-additivity of U, all these coefficients for $s < n$ need to be zero, and that of $U(\mathcal{N})$ (i.e., $s = n$) needs to be 1. So, by induction,

$$p(s,n) = \frac{s}{n-s}p(s-1,n) = \frac{s!}{(n-s)(n-(s-1))\cdots n}p(0,n) = \frac{s!(n-s-1)!}{n!}p(0,n),$$

and $p(n,n) = 1$ implies $p(0,n) = 1$.

The Shapley partition is said to be "fair" because it satisfies the following four properties, again for a super-additive U:

- A player's Shapley value is at least equal to that if it acted alone, i.e., $v_i \geq U(\{i\})$ for all i.

- For two different players i and j, if $U(S \cap \{i\}) = U(S \cap \{j\})$ for all $S \subset \mathcal{N}\setminus\{i,j\}$, then $v_i = v_j$.

- $v_i = 0$ for all *null* players i, i.e., for all players i such that $U(S \cap \{i\}) = U(S)$ for all $S \subset \mathcal{N}\setminus\{i\}$.

- For two different non-negative super-additive U, V gain functions, $v_i^{(U+V)} = v_i^{(U)} + v_i^{(V)}$ for all i.

The last three properties follow by its additive form and its choice of summand, i.e., $U(S \cup \{i\}) - U(S)$. The first property, i.e., that individual players have incentives to join coalitions, follows by lower bounding this summand by $U(\{i\})$.

Again, when Shapley values are computable, they can be used to fairly attribute responsibility to individual players for the communal payoff. For a simple example, suppose we have a system of four players: two content providers, CP1 and CP2, and two competing Internet service providers, ISP1 and ISP2, i.e., $\mathcal{N} = \{ISP1, ISP2, CP1, CP2\}$. Without at least one ISP and one CP, the communal payoff is zero. If either CP1 or CP2 participate, but not both, the payoff is $U_1 > 0$. If both CPs participate, the payoff is $U_2 \geq 2U_1$, as may be the case when the CPs offer different types of content, i.e., $U(\mathcal{N}) = U_2$, $U(\{ISP1, CP1\}) = U_1$, $U(\{CP2\}) = 0$, etc. The Shapley values when all four players participate are, for $k = 1, 2$,

$$
\begin{aligned}
v_{ISPk}(\mathcal{N}) &= \frac{1}{6}U_1 + \frac{1}{12}U_2 \\
v_{CPk}(\mathcal{N}) &= -\frac{1}{6}U_1 + \frac{5}{12}U_2.
\end{aligned}
$$

Note that the sum of the Shapley values of all four players equals U_2, i.e., the communal payoff when all four players participate.

If the content providers competed (offered common content), then it might be the case that $U_2 < 2U_1$, i.e., the payoff is not super-additive (take the case of $\{ISP1, CP1\}$ union $\{ISP2, CP2\}$). If just one ISP were available, i.e., $\mathcal{N} = \{ISP, CP1, CP2\}$, then we could consider competition among the CPs in this simple model (i.e., we could have $U_1 \leq U_2 \leq 2U_1$), however the ISP would now be crucial (i.e., it would have a monopoly). In this case the Shapley values are

$$
\begin{aligned}
v_{ISP}(\mathcal{N}) &= \frac{1}{3}U_1 + \frac{1}{3}U_2 \\
v_{CPk}(\mathcal{N}) &= -\frac{1}{6}U_1 + \frac{1}{3}U_2 \text{ for } k = 1, 2,
\end{aligned}
$$

i.e., the ISP takes a larger share here than *both* ISPs did in the previous (ISP competition) case.

3.4   SUMMARY

In this brief background chapter on games, we considered examples of purely selfish, non-cooperative behavior and altruistic actions. Stability issues were discussed for Nash equilibrium points. Also, games were described involving cooperative players modeled either with individual or communal payoffs, and how to attribute the latter.

Many other potentially useful concepts in game theory have been extensively studied. For example, in a non-cooperative Stackelberg game model, a leader player's action takes place before that of a follower player. For a coalitional game, the *core* is the set of payoff vectors $\underline{v}$ satisfying (3.15) and $U(S) \leq \sum_{i \in S} v_i$ for all $S \subset \mathcal{N}$. So, under a core payoff vector, any coalition $S \subset \mathcal{N}$ is incentivized to remain part of the grand coalition $\mathcal{N}$.

PART II

Peer-to-Peer Search

CHAPTER 4

Search in structured networks

Although the social networks we are interested in may have a very small geodesic diameter, i.e., less than 10 hops, finding short paths remains a difficult problem. In this chapter, we continue certain elements of discussion of caching from Sec .1.3 with a focus on structured search leveraging an explicit addressing system.

This chapter assumes name resolution. The target of an anycast search is to find a peer that, e.g., either possesses certain social attributes, can aid in the acquisition of a certain data object, or can perform a certain service. The specification of the query may involve keywords. A centralized search engine can deal with problems of keywords that have to do with spelling, language translation, even synonyms. Implementing such search functions in a fully distributed peer-to-peer system may result in excessive overhead. A standardized set of keywords for search purposes could be set-up and maintained on behalf of the whole p2p community. Caching information in standard keyword terms, about his friends/neighbors and possibly other peers gleaned from past queries forwarded, would greatly aid the common peer when forwarding future queries. An alternative, hierarchical approach without standardized keywords is a partially centralized system of super-peer index servers, each of which could act as mini search engine on behalf of a local community of peers. This system was used, e.g., by Kazaa clients for file-sharing over FastTrack.

4.1 DISTRIBUTED HASH TABLES (DHTS)

From Sec .1.3, recall the use of consistent hashing to deal with churn (here, *peer* arrivals and departures). In a system using distributed hash tables (DHTs), the identity[1] or *value i* of a peer or object is assigned/bound to a *key*, $x = h(i)$. Any peer i is typically expected to be aware of the value-key bindings $(j, h(j))$ of all objects that are proximal in key space, i.e., for all j such that $h(j) \approx h(i)$. In addition, a peer may maintain of cache of such active bindings elsewhere in key space.

Suppose a peer launches a query for a target with known key, q. In a greedy fashion, a query is simply forwarded to the peer i with *closest* known key $h(i)$ to q. Forwarding continues until the *root* of q, i.e., the closest peer-occupied key to the query's targeted key, is found or the TTL of the query expires. Again, a peer is expected to be able to "resolve" all keys q for which it is the root. For example, if the query objective is for a data object with key q, then its root peer should either possess the data object or know the identity of another peer (or server) that does.

A load-balancing goal of hashing h is that the resulting "peer-occupied" keys are uniformly distributed in the key space.

---

[1]This identity may be further resolved to an IP address of the Internet underlay.

To gauge distances between keys in the Kademlia DHT system [103], the 128-bit key space is equipped with a bitwise exclusive OR (XOR) metric

$$\rho(x, y) \;=\; x \oplus y.$$

If we say two keys x, y are close if $\rho(x, y) \leq 2^{128-k}$ for some fixed $k < 128$, then each peer with key x (i.e., with identity $h^{-1}(x)$) is aware of all peers whose keys share with x its k-bit prefix (k most significant bits). So in the *forwarding neighborhood* of each peer, a region of 2^{128-k} keys, about $v := n2^{-k}$ peers reside if peers are uniformly distributed in key space, where n is the total number of peers in the system. So, if $v \approx \log_2 n$ then the maximum number of hops of a query will be small, i.e.,

$$\frac{128}{k} \;\approx\; \frac{128}{\log_2 n - \log_2(\log_2 n)}$$
$$\approx\; 7,$$

where we used the example total peer population of $n = 10^6$ for the last approximation. In the following section, rather than assuming a certain uniform concentration of peers, other topological assumptions are made consistent with a small worlds graph.

4.2   A SMALL WORLDS CASE

In this section, assume we have a neighborhood topology for a structured keying system wherein each peer has one randomly chosen "long-distance" neighbor of a certain type considered in [86]. A basic result of [86] for such a neighborhood topology shows efficient search performance as a function of the total number of peers in the system. This result was rederived in [12] (Theorem 12) employing Theorem 14 of [81] (Lemma 1 of [12]); see also [80]. A version of these results are described in this section. We first begin with the Lemma attributed to Karp in [12].

Lemma 4.1   *If $X_0 > 1$ and*

$$\mu(z) \;:=\; \mathsf{E}(X_k - X_{k+1} \mid X_k = z) \geq 0$$

is nondecreasing (and not dependent on index k), the number of iterations $T(X_0)$ needed for a nondecreasing, real-valued Markov process $X_0, X_1, X_2, \dots$ to reach 1 is bounded in mean by

$$\mathsf{E} \int_1^{X_0} \frac{1}{\mu(z)} dz \;\geq\; \mathsf{E}T(X_0).$$

Note: By this notation, $X_{T(X_0)} \leq 1$ almost surely for all $X_0 \geq 1$. For an iterative algorithm, $X_k - X_{k-1}$ is the work done by the kth iteration and X_k is the remaining work to be done after the kth iteration.

Proof (sketch [121]): Since $\mu(\cdot)$ is nondecreasing,

$$\int_{X_1}^{X_0} \frac{1}{\mu(z)} dz \geq \frac{1}{\mu(X_0)}(X_0 - X_1) \quad \text{a.s.}$$

So, by taking expectation of both sides we get

$$
\begin{aligned}
E \int_{X_1}^{X_0} \frac{1}{\mu(z)} dz &\geq E \frac{1}{\mu(X_0)}(X_0 - X_1) \\
&= E(E(\frac{1}{\mu(X_0)}(X_0 - X_1) \mid X_0)) \\
&= E(\frac{1}{\mu(X_0)} E(X_0 - X_1 \mid X_0)) \\
&= 1 \\
&= E(T(X_0) - T(X_1)),
\end{aligned}
$$

where the last equality is simply the definition of T. The lemma now follows from an inductive argument.

□

Consider $n \gg 1$ peers (points) residing on a circle with unit radius, recall Figure 1.3. Assume they are evenly spaced ([78, 87] and see Sec .4.3 below), i.e., the (shortest) arc length between nearest neighbors is

$$\varepsilon := 1/n.$$

Each peer can forward messages to either of his nearest neighbors. Also, each peer x can forward messages to $\nu \geq 1$ long-distance peers $L_i(x)$, $1 \leq i \leq \nu$. Let $D(m) := d(x(m), y) \leq 0.5$ be the *shortest* arc length (in any direction) between peer $x(m)$ and data object y on the circle. Assuming that peers are aware of the position in the circle to which they want to send a query, a query from peer $x(m)$ for y is forwarded to the peer $x(m+1)$ in its forwarding table that is closest to y, where peer $x(0)$ originally launched the query for y.

Claim 4.2   If there is a single ($\nu = 1$) long-distance peer in each peer's forwarding table whose distance from the forwarding peer is distributed as $f(d) \propto 1/d$, then the mean search time is $O((\log n)^2)$.

Proof for the case[2] $z \leq 1/4$:

$$
\begin{aligned}
\mu(z) &:= E(D(0) - D(1) \mid D(0) = z) \\
&= \varepsilon \left[\frac{1}{2} + \int_{2z-\varepsilon}^{0.5} f(x) dx \right] \\
&\quad + \int_{\varepsilon}^{z} x f(x) dx + \int_{z}^{2z-\varepsilon} (z - (x-z)) f(x) dx,
\end{aligned}
\tag{4.1}
$$

[2]There is a similar expression for $\mu(z)$ for the other case: $1/4 < z \leq 1/2$.

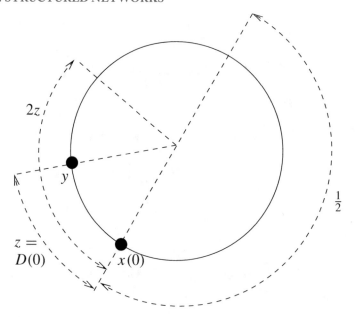

Figure 4.1: A simplified structured search framework for Claim.4.2

where

$$\int_{\varepsilon}^{0.5} f(x)\mathrm{d}x = 0.5. \tag{4.2}$$

Thus,

$$\mu(z) \;\geq\; \int_{\varepsilon}^{z} xf(x)\mathrm{d}x. \tag{4.3}$$

Now since $f(x) = A/x$ for some constant A, we get from (4.2) that

$$A \;=\; -1/(2\log 2\varepsilon) \;=\; 1/(2\log n/2),$$

so that, by (4.3),

$$\mu(z) \;\geq\; (z - \varepsilon)A.$$

By direct differentiation of (4.1), $\mathrm{d}\mu(z)/\mathrm{d}z \geq 0$.

So, by Lemma 4.1, the mean search time is bounded above by

$$
\begin{aligned}
\int_{2\varepsilon}^{0.25} \frac{1}{\mu(z)} dz \;\; &\leq \;\; \int_{2\varepsilon}^{0.25} \frac{1}{(z-\varepsilon)A} dz \\
&= \;\; O\!\left(\log n \int_{2\varepsilon}^{0.5} \frac{1}{z-\varepsilon} dz \right) \\
&= \;\; O((\log n)^2) \,.
\end{aligned}
$$

☐

In [86], a key space of dimension $n \geq 2$ (Cartesian coordinates) was considered under the L_1 norm with $f(x) = A/x^n$.

4.3   LOAD BALANCING

In the previous section, we assumed evenly spaced peers in key space. As at the end of Sec .4.1, consider n peers whose keys are chosen independently and uniformly at random on a circle of unit circumference and assign to each peer the arc segment for which its key is the left boundary point. Let A_n be a random variable representing lengths of arc segments[3]. The goal is to avoid long arcs since this corresponds to proportionately greater load on the peers responsible for them.

Lemma 4.3

$$
EA_n \;\; = \;\; \frac{1}{n} \approx \sigma(A_n),
$$

where $\sigma(A_n)$ is the standard deviation of A_n and the approximation $(\approx)$ is close for large n.

Proof. The error function is

$$
P(A_n > x) \;\; = \;\; (1-x)^{n-1}.
$$

One can easily check that

$$
EA_n \;\; = \;\; \int_0^1 P(A_n > x) dx = \frac{1}{n}.
$$

To compute the standard deviation $\sigma(A_n)$, one can compute the PDF of A_n,

$$
-\frac{d}{dx} P(A_n > x),
$$

[3]The arc segment's length is obviously proportional to the number of keys resident on it.

and then directly evaluate EA_n^2 (integrate by parts twice) and substitute into $\mathrm{var}(A_n) := \sigma^2(A_n) := EA_n^2 - (EA_n)^2$.

□

To reduce the variance of assigned arc length in key space, suppose that each peer has k keys assigned, where all nk keys are chosen and arc segments are assigned as above. Let $B_{n,k}$ be a random variable representing the total lengths of the k arc segments of a peer. Clearly, by the previous lemma,

$$EB_{n,k} = kEA_{nk} = \frac{1}{n} = EA_n,$$

i.e., the mean total arc length assigned to a peer does not depend on k.

Claim 4.4   For large n,

$$\sigma(B_{n,k}) \approx \frac{\sigma(A_n)}{\sqrt{k}}.$$

Proof. Let $N = nk$ and pick k distinct indices from $\{1, \ldots, N\}$ at random, independently of the original experiment, and let B be the sum of the lengths corresponding to these indices. More precisely, let $(\xi_1, \ldots, \xi_k)$ be a sampling without replacement from $\{1, \ldots, nk\}$ and let

$$B = A_{\xi_1} + \cdots + A_{\xi_k}.$$

Observation 1: The random vector $(A_1, \ldots, A_N)$ is permutable, and so is $(\xi_1, \ldots, \xi_k)$. Therefore,

$$B \sim A_1 + \cdots + A_k.$$

Observation 2: Let $X_1, X_2, \ldots$ be i.i.d. exponentials with mean 1, and let $S_n = X_1 + \cdots + X_n$. Then,

$$(A_1, \ldots, A_N) \sim (X_1, \ldots X_N | S_N = 1)$$

by conditional uniformity [157]. From this one can directly compute $E(A_1 + \cdots + A_k | S_N)$ and $\mathrm{var}(A_1 + \cdots + A_k | S_N)$ and then evaluate at $S_N = 1$. To compute the density f_B of B, note that

$$B \sim (S_k | S_N = 1),$$

implies

$$P(B \in dx) = \left. \frac{P(S_k \in dx, S_N \in dt)}{P(S_N \in dt)} \right|_{t=1}.$$

Let $f_m(x)$ be the density of S_m (which is gamma):

$$f_m(x) = \frac{x^{m-1}}{(m-1)!} e^{-x}.$$

Hence,

$$f_B(x) = \frac{f_k(x) f_{N-k}(1-x)}{f_N(1)} \propto x^{k-1}(1-x)^{(n-1)k-1}$$

which is beta distributed with parameters

$$\alpha = k \quad \text{and} \quad \beta = (n-1)k.$$

According to [155],

$$\mathsf{E}B = \frac{\alpha}{\alpha+\beta} \quad \text{and} \quad \mathrm{var}(B) = \frac{\alpha\beta}{(\alpha+\beta)^2(\alpha+\beta+1)}.$$

So $\mathsf{E}B = k/N = 1/n$, and

$$\mathrm{var}(B) = \frac{n-1}{n^2(nk+1)} \approx \frac{1}{n^2k} \text{ as } n \to \infty.$$

☐

Thus, higher k reduces the variance of total arc assigned to each peer but increases the overhead associated with multiple assigned keys. By Chebyshev's inequality, $\mathsf{P}(|B - \frac{1}{n}| > \frac{1}{n}) \leq n^2\mathrm{var}(B) \approx \frac{1}{k}$. The Chord DHT suggests $k = \mathrm{O}(\log n)$ [78].

4.4   DISCUSSION

To set up peer neighborhoods as described above, active peers are expected to collect information (value-key bindings) about each other. This can be actively done by querying known individual longer-lived (super) peers or by efficient local broadcasting in key space using limited-scope flooding. The scope of the broadcast/flood can be limited in breadth (how many peers are simultaneously contacted by a single relaying peer) and limited in depth (by TTL). Also, redundant relaying can be mitigated by the use of unique flood identifiers.

Peers can also learn topological information passively through reverse-path response [40] to relayed queries, i.e., in the case of a successful query, the resolving peer acknowledges back along the path the query took (it may directly communicate the requested information to querying peer or merge it with a reverse-path acknowledgement). Such continual network monitoring can determine which peers are presently logged-on and active, and can identify the more reliable longer-lived peers in particular.

In the framework of Sec .4.1, Kademlia's metric is unidirectional in that all paths for any given query target will converge. Peers may cache responses to queries of popular target keys of which they are not roots using the techniques just mentioned for topological discovery. Such non-root caching may alleviate querying congestion in the key space near a popular target object, i.e., congestion due to a flash crowd.

The effects of peer churn can be mitigated by creating a "multihomed" system for queries wherein any particular key has multiple root peers, i.e., proactive redundancy. Multihoming creates more resolving load per peer, of course, and requires that the root peers of any given key communicate so that they are consistent about how to resolve it. Detection of erroneous or trojan data objects can be based on obtaining ostensibly the same data object from two or more of its roots.

4.5   FORWARDING WITH PRIVACY

In this section, we discuss a basic method for forwarding with anonymity. These techniques can also be applied in an unstructured setting, *cf.* next chapter.

BLIND PROXYING

If peer A wishes to contact B in an anonymous fashion, it may send a message to a relaying entity R, the message being addressed to B with return address A. Relay R replaces the address of A in the message with its own address before forwarding to B. R would need to track which particular sessions $A \leftrightarrow R$ and $R \leftrightarrow B$ to splice. The response from B would be transmitted to A again via R (i.e., reverse-path response). Thus, if an entity intercepts the message between R and B it will not be able to identify that the correspondent with B is A. If the application involved is an HTTP download from client A and (web) server B, the TCP session is effectively segmented at R.

It may be possible that the message is intercepted at a point between A and R. To protect against this possibility, the message can be dispatched from A with destination address R *and* the destination B encrypted with the public key of R. Also, the message content itself (the URL or corresponding data object) could be encrypted by public key (of the requested destination, B, or the requesting source, A, respectively).

Obviously, the message content can also be encrypted by A using the public key of B.

ONION ROUTING

Tor is presently the most popular anonymity framework on the Internet. Any host can join as a client and volunteer to relay. A small group of trusted, authoritative servers maintain a directory of the nominally reliable, relaying hosts. Hosts ascertain their relays simply by contacting a relay directory server. Thus, Tor provides an encrypted circuit from client host to (or from) entry relay node to intermediate relay node to exit relay node, i.e., at least three relays per encrypted circuit. The Tor exit relay node communicates directly with, e.g., public web servers. A relaying party only knows the identities of its two immediate neighboring parties in the circuit, e.g., the intermediate relay knows the identities (IP addresses) only of the entry and exit relay, and the exit relay knows the identities of the servers and intermediate relay. These three relays are iteratively identified by the client, starting with the entry relay, and cryptographic key exchanges occur to establish secure connections within the circuit.

The overhead of this relay directory system suffers quadratic cost in the number of client hosts. Every client must periodically download a list of all relays, so if there are n client hosts and r relays then the total overhead to ascertain relays is $O(nr)$. If the number of relays grows proportionately with the number of clients, i.e., $r = cn$ for some constant c, then this cost is $O(n^2)$.

In [105], an alternative to the centralized registry system of Tor is described. A DHT system is proposed to better scale and protect the identity of the relaying peers, even in the presence of churn (arrivals and departures of relays). First recall the Kademlia DHT system briefly described in Sec .4.1. A client host would select a relay by choosing a key k at random and querying for the relay whose key is closest to k, i.e., the (relay) root of k.

To authenticate a putative root for k [149], a querying peer would, by local flooding, request corroboration by all peers that reside in a neighborhood of k in key space. Neighborhood consensus could be required before a peer is accepted as the root. This consensus would require a secure management protocol to deal with peer churn, i.e., peer arrivals (neighbors dynamically issue authenticatable certificates to attest to the new peer) and departures (certificates are revoked). Such a system would prevent a peer from falsely claiming to be a root and prevent deliberate mis-routing of a query.

The peer-to-peer system Freenet [35, 165] uses both limited-scope flooding and reverse-path response to find *and* distribute data objects. Under Freenet, the initial query TTL can be varied and intermediate Freenet nodes cache the data objects themselves. Both techniques can be used to obfuscate the identity (in this way protect the privacy) of the query originator. Finally, we mention I2P which is a related, secure anonymity framework available to applications, such as file-sharing and messaging.

CHAPTER 5

Search in unstructured networks

As with the previous chapter, this one is concerned with the problem of search for a peer that possesses certain attributes or content, that can perform a service, or otherwise belongs to a social clique. In an unstructured social network, a clique is born and can be discovered by fully distributed random-walk and limited-scope flooding techniques[1], alone or in combination, e.g., [144]. Through querying and reverse-path forwarding of query responses, peers can learn that a particular clique is sufficiently permanent (popular and either large or churn-free) so that it is worth caching information about it for their own purposes, or for purposes of query resolution for other peers. Also, peers learn about certain others who are very reliable, honest, cooperative, and well-connected; partially centralized search systems rely on such hubs/super-peers. The material in this chapter is covered in greater depth by, e.g., [4, 47, 63, 64, 68, 97, 108].

5.1 SINGLE THREADED SEARCH BY RANDOM WALK

Consider Boolean unidirectional edge weights $w_{i,j} \in \{0, 1\}$ indicating whether an edge exists from i to j. Again, d_i is defined to be the out-degree of i, $d_i = \sum_{j \neq i} w_{i,j}$. A random walk is governed by the Markovian transition rate probabilities,

$$P_{i,j} \;=\; \frac{w_{i,j}}{d_i} \quad \text{for all } j \neq i,$$

i.e., $P_{i,j}$ is the probability that i forwards a query to j. Note that the transition probability matrix $\mathbf{P}$ (with elements $P_{i,j}$) is *row stochastic*, i.e., $P_{i,j} \geq 0$ for all i, j and $\sum_j P_{i,j} = 1$ for all i, where we sum over the $n := |V|$ vertices.

If $\mathbf{P}$ is irreducible[2] and aperiodic[3], then there is a unique $\underline{\pi} \in \Sigma_n$ (the n-dimensional simplex) such that all rows of

$$\mathbf{P}^\infty \;:=\; \lim_{k \to \infty} \mathbf{P}^k$$

are $\underline{\pi}^{\mathrm{T}}$, where $\underline{\pi}$ is the unique stationary invariant distribution of the discrete-time Markov chain with transition probability matrix $\mathbf{P}$: $\underline{\pi}^{\mathrm{T}} = \underline{\pi}^{\mathrm{T}}\mathbf{P}$, see Ch. 6 of [67]. In our case, $\pi_i = d_i/D$, where $D = \sum_i d_i = nd/2$ is the total number of edges in the graph of n vertices (peers). Here, the detailed balance equations hold:

$$\pi_i P_{ij} \;=\; \pi_j P_{ji} \in \{\tfrac{1}{D}, 0\} \quad \text{for all } i, j,$$

---

[1]The Gnutella file-sharing system used limited-scope flooding for unstructured search, recall Sec .4.4.
[2]That is, at least one unidirectional path exists between any given peer to any other, implying the social graph is connected.
[3]For all peers, the (geodesic) length in hops of all loop-free return paths have greatest common divisor equal to one.

i.e., the stationary Markov chain with transition probability matrix $\mathbf{P}$ is *time-reversible*. To reiterate, these assumptions are an idealization from a social networking context considering how social ties generally lack the symmetry and transitive properties.

The above assumptions imply that all of the eigenvalues of $\mathbf{P}$ are real and the set of eigenvectors are orthogonal[4]. Let Λ be the second largest magnitude of an eigenvalue of $\mathbf{P}$, where $\Lambda < 1$ since $\mathbf{P}$ is a row-stochastic matrix [71][5]. More specifically, Λ is the maximum of the second largest eigenvalue $\lambda_2 < 1$ and the absolute value of the smallest eigenvalue λ_n, recalling that $\mathbf{P}$ is $n \times n$.

The time to convergence of the random walk to its stationary distribution is governed by Λ. We now express Λ in terms of graphical attributes that are meaningful from a social networking point of view; see [45, 47, 64].

Claim 5.1

$$\Lambda \;\geq\; \max_{S \subset V} |H(S)|$$

where H is the group homophily measure given in (2.8).

Proof. For all real vectors $\underline{v}^* \neq \underline{0}$ [47, 71]:

$$-1 < \lambda_n = \inf_{\underline{v} \neq \underline{0}} \frac{\langle \mathbf{P}\underline{v}, \underline{v}\rangle_{\underline{\pi}}}{\langle \underline{v}, \underline{v}\rangle_{\underline{\pi}}} \leq \frac{\langle \mathbf{P}\underline{v}^*, \underline{v}^*\rangle_{\underline{\pi}}}{\langle \underline{v}^*, \underline{v}^*\rangle_{\underline{\pi}}} \leq \lambda_2 = \sup_{\underline{v} \neq \underline{0}, \, <\underline{v}, \underline{1}>_{\underline{\pi}}=0} \frac{\langle \mathbf{P}\underline{v}, \underline{v}\rangle_{\underline{\pi}}}{\langle \underline{v}, \underline{v}\rangle_{\underline{\pi}}} < 1 \,,$$

where the inner product $\langle \underline{u}, \underline{v}\rangle_{\underline{\pi}} := \sum_i u_i v_i \pi_i$, and the constraint in the expression for λ_2 is orthogonal eigenvectors.

Since $\Lambda = \max\{|\lambda_n|, \lambda_2\}$, it will suffice to find a non-zero real vector $\underline{v}^*(S)$ such that $< \underline{v}^*(S), \underline{1} >_{\underline{\pi}} = 0$ and

$$\frac{\langle \mathbf{P}\underline{v}^*(S), \underline{v}^*(S)\rangle_{\underline{\pi}}}{\langle \underline{v}^*(S), \underline{v}^*(S)\rangle_{\underline{\pi}}} \;=\; |H(S)|.$$

Such a vector is

$$v^*(S) \;=\; \begin{cases} (|S|d_i)^{-1} & \text{if } i \in S \\ ((n - |S|)d_i)^{-1} & \text{if } i \notin S \end{cases}$$

$\square$

---

[4] $\tilde{P}_{ij} := \sqrt{\pi_i}\, P_{ij}/\sqrt{\pi_j}$ is self-adjoint (Hermitian), i.e., $\tilde{\mathbf{P}} = \tilde{\mathbf{P}}^\mathrm{T}$. So, the Courant-Fischer-Weyl min-max variational spectral characterization applies to $\tilde{\mathbf{P}}$ [71]. Now note that $\tilde{\mathbf{P}} = \mathbf{Q}\mathbf{P}\mathbf{Q}^{-1}$ where $\mathbf{Q} = \mathrm{diag}(\sqrt{\pi_1}, \sqrt{\pi_2}, ..., \sqrt{\pi_N})$, i.e., $\tilde{\mathbf{P}}$ is similar to $\mathbf{P}$. Therefore, these matrices have a common spectrum (set of eigenvalues), where a left eigenvector $\underline{v}^\mathrm{T}\mathbf{Q}^{-1}$ of $\tilde{\mathbf{P}}$ corresponds to the left eigenvector $\underline{v}^\mathrm{T}$ of $\mathbf{P}$.

[5] The Perron-Frobenius theorem figures here. For the eigenvalue of largest magnitude, $\lambda_1 = 1$, $\mathbf{P}$ has corresponding left eigenvector $\underline{\pi}$ and right eigenvector $\underline{1}$ (both with all elements non-negative).

There are other similar results showing how the parameter governing convergence speed (or mixing time), Λ, is related to the problem of finding a small number of edges which, when removed, would disconnect a large number of vertices, see e.g., [45, 68] and *cf.* Ch. 7. Again in practice, search performance would improve if multiple random walks were conducted simultaneously, i.e., greater breadth of search, at the expense of additional overhead of course. Time-to-live (TTL) mechanisms can limit the depth of the search of each thread (random walk). Also, to avoid redundant forwarding decisions, each thread can employ a taboo list and/or the forwarding vertices can store identifiers for each search process.

5.2   POPULARITY BIASED RANDOM WALKS

Consider a graph of bidirectional edges with Boolean weights. Given the edge degrees (d_i for vertex i), suppose by random walk we wish to achieve a particular stationary distribution on the vertices, q. For example, recalling the "square-root" results of Sec .1.3 with objective given by Equation (1.5), we may desire $q_i \propto p_i$, where p_i is the *popularity* of (the content in cache) i [166]. As in a Metropolis-Hastings approach, the transition probability matrix $\mathbf{P}$ defined above is adjusted so that the invariant changes to q from $\underline{\pi}$, where

$$\pi_i = d_i \left/ \sum_j d_j \right. .$$

Claim 5.2   If $\mathbf{P}$ is time-reversible, then the transition probability matrix $\mathbf{P}^{(q)}$ with elements

$$P_{i,j}^{(q)} = \begin{cases} \frac{1}{2} P_{i,j} & \text{if } \frac{\sqrt{p_i}}{d_i} \leq \frac{\sqrt{p_j}}{d_j} \\ \frac{\sqrt{p_j}}{2\sqrt{p_i}} P_{i,j} & \text{if } \frac{\sqrt{p_i}}{d_i} > \frac{\sqrt{p_j}}{d_j} \end{cases}$$

has invariant q with elements

$$q_i = \frac{\sqrt{p_i}}{\sum_j \sqrt{p_j}}.$$

Proof. Note that $\mathbf{P}^{(q)}$ is also time-reversible and check that q satisfies its detailed balance equations. $\square$

As an example, suppose the active peers learn about a small subset of $k < n$ super-peers (reliable, powerful, long-lived, generally cooperative peers). To bias search toward super-peers, consider parameters $p^{(h)} \gg p^{(l)} > 0$ such that $q_i \propto p^{(h)}$ if i is a super-peer, else $q_i \propto p^{(l)}$.

5.3   DISCUSSION

The rate of convergence of the empirical distribution of the ergodic Markov random walk to its invariant distribution has also been explored by large deviations (e.g., Theorem 3.1.6 of [46]). In the Markov Chain Monte Carlo literature, e.g., [27], a variety of techniques have been proposed to dynamically improve the performance of single random-walk threads. When social groups can communicate through shared peers [116, 118, 151], learned membership correlations can be used for anycast search, e.g., [121].

PART III

Reputations and Referrals

CHAPTER 6

Transactions, reputations, and referrals

In this chapter, we introduce a simple model of distributed social referral. Referrals are generally based on non-transitive and non-symmetrical reputations between individuals. Significant reputation for purposes of referral, sometimes called social capital, is often earned and accrued through past transactions online or exists because of some *a priori* social association such as a family tie[1]. We will focus on accrued reputations which will be taken to be non-negative real-valued features of a unidirectional edge between two peers. That is, the weight of the edge (k, i),

$$R_{ki}^{(d)} \quad := \quad w_{k,i}$$

in a referral social graph is the (direct) reputation of k from the point-of-view of i. Reputations[2] and referrals based on them can impact whether and how transactions take place between peers. That is, past transactional activity can impact present and future transactions through a reputation-referral system. If properly designed, a reputation-referral system can incentivize certain cooperative behavior among peers, *cf.* Ch. 8. A far from completely connected social graph is used for referrals, although transactions can potentially occur between any two peers. Distributed referral systems may hierarchically pool referral information to deal with issues of scale and calibration. In such systems, care should be taken to limit the ability of one or more referring peers to unfairly defame or falsely hype, *cf.* Ch. 7.

6.1 ACCRUED REPUTATIONS

Assume $n + 1$ interacting peers all of whom can directly transact with each other over the Internet. Also, assume all peers operate honestly and fault-free. In the following, a *transaction* directly involves a pair of peers and is assumed for simplicity here to have a Boolean outcome. If a transaction outcome involving i and j was good from i's point-of-view, then i could update its direct reputation values

---

[1] Less significant reputations include an initial benefit of the doubt for a newly joined member of a social group which can be used to engage in some transactions within the group and thereby begin to accrue a more significant direct reputation from the point of view of its other members.

[2] The direct reputation, an edge attribute in the social graph, is sometimes called "trust", and the average direct reputation of a vertex is sometimes simply called its "reputation", e.g., [76].

according to

$$R_{ki}^{(d)} \rightarrow \begin{cases} (R_{ji}^{(d)} + C)/(\sum_l R_{li}^{(d)} + C) & \text{if } k = j \\ R_{ki}^{(d)}/(\sum_l R_{li}^{(d)} + C) & \text{else} \end{cases} \qquad (6.1)$$

for a fixed reward parameter $C > 0$, i.e., j's reputation is rewarded by the amount C. Note that at each peer i, the reputations of others are assumed always normalized, i.e., $\sum_{k \neq i} R_{ki}^{(d)} = 1$, and how bad outcomes need not be penalized as normalization will naturally "age-out" reputations. Also note that there is a simple trade-off in the chosen value of the reward parameter: if C is too high then the peers may derive too much reward from too few transactions, and similarly not enough reward if C is too low.

Assume that a peer j's transaction response does not depend on the requesting peer i; thus, we denote with a single subscript ε_j as peer j's *propensity to cooperate* in *unilateral* transactions with Boolean outcomes[3]. If all transaction pairs (i, j) are equally likely, then essentially by the law of large numbers the direct reputations (6.1) satisfy [112]: for all k, j, i,

$$R_{ki}^{(d)}/R_{ji}^{(d)} \rightarrow \varepsilon_k/\varepsilon_j \qquad (6.2)$$

as the number of transactions grows. That is, the reputations reveal the propensities to cooperate.

In variations of the above framework, normalization could employ an estimate of the current transaction rates: $\rho_{i,j}$ is the fraction of i's transactions that involve peer j, and for each new transaction with peer j update

$$\rho_{i,k} \rightarrow \begin{cases} (\rho_{i,j} + 1)/(1 + \sum_{k' \in N_i} \rho_{i,k'}) & \text{if } k = j \\ \rho_{i,k}/(1 + \sum_{k' \in N_i} \rho_{i,k'}) & \text{else} \end{cases}$$

where N_i is the set of peers j for which $\rho_{i,j}$ is significantly greater than zero[4]. Also, one could have transactions with partial success so that the reward could range in $[0, C]$, or one could explicitly penalize bad transaction outcomes, i.e., in (6.1) use $(\ldots - C)^+$ instead of $(\ldots + C)$.

6.2   REFERRALS IN DISTRIBUTED PEER-TO-PEER SYSTEMS

Suppose indirect referrals will be used by a peer i who wishes to obtain reputation information about j even though i and j have had no or insufficient prior direct contact, or direct contact occurred in the distant past and has been forgotten. A general property of the referral value of a path from j to i in social graph is that it is non-decreasing in the component edges' direct reputation values of the referring peers in the path. In the following framework, edge reputations are multiplied along the path and the referral results of different paths are added. That is, the system considered here will be specified by matrix multiplication, similar to that of, e.g., [32, 76].

---

[3]On the other hand, in bilateral BitTorrent i, j transactions, from peer i's point of view, the quality of a peer j's response, i.e., j's allocated uplink bandwidth, depends on i's uplink, *cf.* Ch. 8.

[4]In practice, for systems with very large peer populations, obviously peers can only afford to cache reputations which are significant. The case of no prior contact may be flagged so that zero reputation clearly indicates significantly poor transaction outcomes.

ONE-STEP REFERRALS

For i to indirectly acquire reputation information about j, define

$$J_{ji}^{(1)} = \sum_{k \neq j,i} R_{jk}^{(d)} R_{ki}^{(d)}, \tag{6.3}$$

where a "correct" (honest and not faulty) peer k will refer j using their own direct reputation values. So, the referred reputations are *weighted* by the direct reputations $R_{ki}^{(d)}$ of the referring peers k and *self referrals* are excluded, i.e., $k \neq j$.

Referrals may be used to augment direct reputation values. For example, this can be done additively:

$$R_{ji}^{(1)} = a J_{ji}^{(1)} + (1-a) R_{ji}^{(d)} \tag{6.4}$$

for some non-negative constant $a \leq 1$. Referrals about j from highly reputed peers generally could have a highly dominant effect on the $J^{(1)}$ component of the first term [76]. Alternatively, referrals could be used only in the absence of direct prior contact:

$$R_{ji}^{(1)} = \begin{cases} R_{ji}^{(d)} & \text{if prior direct contact} \\ J_{ji}^{(1)} & \text{else.} \end{cases} \tag{6.5}$$

MULTIPLE-STEP REFERRALS

The system (6.5) may continue to exchange indirect reputation referrals until direct reputation informs all pairs of peers. For a system with $n + 1$ peers, consider the arbitrary peer labeled $n + 1$. Let $\underline{R}_{n+1}^{(d)}$ be the row n-vector with entries $R_{(n+1)i}^{(d)}$ and let $\mathbf{R}^{(d)}$ be the $n \times n$ zero-diagonal matrix with (k, i) entries $R_{ki}^{(d)}$ for $k, i \leq n$. For $m > 0$, define the m-step referrals

$$\underline{J}_{n+1}^{(m)} = \underline{R}_{n+1}^{(d)} (\mathbf{R}^{(d)})^m. \tag{6.6}$$

For an additive system like (6.4), we can compose

$$R_{ji}^{(m)} = \sum_{l=0}^{m} \alpha_l^{(m)} J_{ji}^{(l)} \quad \text{for all } i \neq j, \tag{6.7}$$

where $J^{(0)} := R^{(d)}$. If $\underline{\alpha}^{(m)} \in \Sigma_{m+1}$, the $(m + 1)$-dimensional simplex, and is of the form $\alpha_l^{(m)} = a^l / A_m$ for $0 < a < 1$ with $A_m := \sum_{l=0}^{m} a^l$, then the referrals (6.7) will converge as $m \to \infty$. For example, for peer $n + 1$,

$$\lim_{m \to \infty} \underline{R}_{n+1}^{(m)} = (1-a) \underline{R}_{n+1}^{(d)} (\mathbf{I} - a\mathbf{R}^{(d)})^{-1}.$$

Note that $A_m \to (1-a)^{-1}$ and $\mathbf{R}^{(d)}$ is a column sub-stochastic matrix[5] [71, 72]. Recall conditions for convergence of Markov chains were discussed in Ch. 5, also *cf.* Ch. 9 for similar consensus dynamics.

Alternatively, if the referrals are simply added together, i.e., $\alpha_l^{(m)} = 1$ for all l, m, then

$$\lim_{m\to\infty} \underline{R}_{n+1}^{(m)} = \underline{R}_{n+1}^{(d)}(\mathbf{I} - \mathbf{R}^{(d)})^{-1}.$$

ADDITIONAL VARIATIONS

Different types of averaging have been considered in the context of gossip propagation[6] [26] (and *cf.* Ch. 10) and for recommender systems [44, 101]. The referrals of the above system could be normalized as in (6.1). The referral value of a path can instead be taken to be the direct reputation of the penultimate peer in the chain (i.e., immediate neighbors of the recipient of the referral); this may be the case in privacy-preserving referral systems, recall Sec .4.5 and see, e.g., [120]. Also, the results of different referral paths can be maximized rather than added. See [141].

6.3   DISCUSSION

Eleven different reputation systems are compared in [133]. In a flat, distributed peer-to-peer implementation of the above system with n peers, each referral iteration obviously requires a total of $O(n^2)$ message-passing overhead (possibly through a kind of masked multicast to prevent peers from receiving direct feedback about referrals regarding themselves). Alternatively, current direct reputation data could be batched and periodically sent to a single trusted super-peer (reputation-referral index server) which could calculate referrals on behalf of the peer community and disseminate them periodically or on-demand. A distributed and scalably hierarchical super-peer (or managed server-based) framework would be based on peer groupings, and inter-group reputation queries could be handled by the super-peers or referral servers. Also, the identities of particularly high or low reputed peers could be more broadly disseminated in the hierarchy in an automatic way. Generally, to reduce the required communication, computation and memory, potential sparseness and/or quantization of the inter-group direct reputations $R^{(d)}$ could be leveraged. Recall reference to Bloom filters [90, 94] in Sec .1.3, a technique which may also serve to reduce communication costs for reputation/referral servers. Proposed distributed referral frameworks can be found in [1, 76, 96, 162].

---

[5]That is, $0 \leq \sum_{k \leq n} R_{ki}^{(d)} = 1 - R_{(n+1)i}^{(d)} \leq 1$.
[6]Also, asynchronous updates and rates of convergence have been considered.

CHAPTER 7

False Referrals

A peer x referring y acts as a relay of either experienced direct reputation of y (i.e., $R_{yx}^{(d)}$) or of received indirect referred reputation information from y about another peer; in the latter case, assume x forwards the referral multiplied by $R_{yx}^{(d)}$ as in the previous chapter. A false referral occurs when x instead forwards a quantity different from that prescribed by the referral framework.

We consider here deliberate false referrals pertaining to "sybil" identities. Suppose a peer has an identity x associated with legitimately accrued reputation. Through self-referral, he wishes to impart this reputation to a group S of sybil identities which he controls. Prior work on *detecting* sybil attackers has to do with determining how certain identities are always clustered together, e.g., physically in a mobile ad-hoc network (MANET) [125] or virtually in a social referral graph [42, 45, 163]. In the latter case, the problem is related to determining whether x is *between* S and the rest of the world, i.e., determining whether the number of edges between $S \cup \{x\}$ and the rest of the world is much more typical of a single peer (x) than of $1 + |S|$ peers for $|S| \gg 1$.

In the following, by basically using a kind of conservation of earned social capital, we show that the referral framework of the previous chapter is robust in the presence of sybil identities.

7.1 SYBILS

A sybil attacker is a peer employing multiple (sybil) identities with malicious or selfish intent [30, 50]. For the BitTorrent file-sharing example (*cf.* next chapter), a swarm's newly arrived peers engage in one-way client-server transactions as they have nothing to swap initially. Social networks may often have such initial "benefit of the doubt" etiquette for newly arrived peers which can be exploited by sybil attackers, i.e., free riding. Sybil activity is also used to discard an identity that was sullied due to malicious or greedy behavior and simply acquire a new one, i.e., white-washing. Finally, sybils are a significant threat to referral systems, i.e., they mask self-referrals used to hype themselves or stage defamation attacks on third parties.

Standard counter-measures to sybil activity include instituting identity acquisition costs and identity authentication overhead (e.g., binding social identities to fixed IP addresses). In the file-sharing example, ahead of client-server transactions the system could require solving a preliminary computational puzzle including those necessarily involving human participation, e.g., CAPTCHA. Also, transmitting only encrypted file segments for client-server transactions would require a decryption task of the client peer, but this particular approach would involve additional storage and computation overhead for all peers. Alternatively, for a file-sharing system, one could distribute only common file segments to newly arrived peers so that sybil identities would gain little more; however, such a system would give segments of little trade value to legitimate newly arrived peers. Again, note that a sybil attacker may engage in legitimate transactions and accrue earned reputation rewards as a result. However, these rewards may naturally be distributed, and hence diluted, among the sybil identities.

In this chapter, we will use a reputation system that accounts for transaction outcomes between the directly participating parties. Some basic assumptions of the referral system are: peer identity authentication (which itself does not preclude sybil identities), secure referral communications over the Internet underlay [43, 162, 163], and synchronous information exchange for simplicity.

Consider two scenarios. The first is that of the previous chapter where indirect reputations are computed when no sybils are present among the $n + 1$ peers. The sybil attack scenario is the same as the first except that the $(n + 1)^{\text{th}}$ peer is assumed to be a peer identity that is part of a sybil group S. Thus, there are $n + |S|$ different peer identities in the system, though only $n + 1 < n + |S|$ distinct peers. We assume that the sybil identities refer each other with maximum reputation R^{max}. In the following, we use tilde '$\sim$' to indicate reputations in the presence of sybil attack.

As mentioned earlier and as assumed in [32], *direct* reputation values are assumed divided among sybil identities. Thus,

$$R^{(\text{d})}_{(n+1)i} = \sum_{k \in S} \tilde{R}^{(\text{d})}_{ki} =: \tilde{R}^{(\text{d})}_{Si} \text{ for all } i \notin S. \tag{7.1}$$

This constraint, which is fundamental for the following developments, requires that strictly positive default initial direct reputations (benefit of the doubt) are not used. Instead, referrals (indirect reputations) need to be used when direct reputations are basically zero. Again, if needs be, a separate flag could be used to indicate whether zero represents no information or poor reputation.

7.2   SYBIL-RESISTANT ONE-STEP REFERRALS

In this section, we consider the effects of sybils on just one-step referrals. We say that a system is *sybil-proof* if there is no sybil strategy that significantly affects reputation information [32]; the effect is a bounded function of the number of sybil identities in particular.

Here, take $n + 1 \in S$ as a typical sybil identity in the attack scenario, whose reputation from the point of view of $i \notin S$ is

$$\tilde{J}^{(1)}_{(n+1)i} := \sum_{k \notin S \cup \{i\}} R^{(\text{d})}_{(n+1)k} R^{(\text{d})}_{ki} + \sum_{k \in S \setminus \{n+1\}} R^{\text{max}} \tilde{R}^{(\text{d})}_{ki}.$$

By (7.1),

$$\tilde{J}^{(1)}_{(n+1)i} = J^{(1)}_{(n+1)i} + R^{\text{max}}(\tilde{R}^{(\text{d})}_{Si} - \tilde{R}^{(\text{d})}_{(n+1)i})$$
$$\leq J^{(1)}_{(n+1)i} + R^{\text{max}} R^{(\text{d})}_{(n+1)i}, \tag{7.2}$$

thus limiting the sybil's self-referral. Note that the quantity $R^{\max}$ in the previous inequality may be smaller in the sybil scenario due to the ostensibly larger number of peers (likewise the $R^{(d)}_{(n+1)i}$ term if the reputations are locally normalized).

Furthermore, for two different peers $j, i \notin S$, assume the sybil attacker wants to refer j to i using $\tilde{R}_{jS}$ to unfairly hype ($\tilde{R}_{jS} = R^{\max}$) or defame ($\tilde{R}_{jS} = R^{\min} \geq 0$) j, i.e., it may be the case that $\tilde{R}_{jS} \neq R^{(d)}_{j(n+1)}$. So, by (7.1), we see

$$
\begin{aligned}
\tilde{J}^{(1)}_{ji} &:= \sum_{k \notin S \cup \{i,j\}} R^{(d)}_{jk} R^{(d)}_{ki} + \sum_{k \in S} \tilde{R}_{jS} \tilde{R}^{(d)}_{ki} \\
&= \sum_{k \notin S \cup \{i,j\}} R^{(d)}_{jk} R^{(d)}_{ki} + \tilde{R}_{jS} R^{(d)}_{(n+1)i} \\
&= J^{(1)}_{ji} + \tilde{R}_{jS} R^{(d)}_{(n+1)i}.
\end{aligned}
\tag{7.3}
$$

Thus, there is little motive for a peer to adopt sybil identities to affect the one-step referrals of another peer. Note that this does not mean that there is *no* effect by simple false referral on the reputation of $j \notin S$, i.e., $\tilde{J}^{(1)}_{ji} \neq J^{(1)}_{ji}$ if $\tilde{R}_{jS} \neq R^{(d)}_{j(n+1)} = R^{(d)}_{jS}$, *cf.* Sec .7.3. So, the extent to which the accrued reputations or one-step referrals (6.5) are sybil-proof is given by (7.1), (7.2) and (7.3).

Claim 7.1   The additive one-step reputation system (6.4) is sybil-proof if

$$
a \leq \frac{1}{R^{\max} + 1}.
\tag{7.4}
$$

Proof. For the sybil attack scenario,

$$
\tilde{R}^{(1)}_{(n+1)i} = a J^{(1)}_{(n+1)i} + a R^{\max}(\tilde{R}^{(d)}_{Si} - \tilde{R}^{(d)}_{(n+1)i}) + (1-a)\tilde{R}^{(d)}_{(n+1)i}
$$

subject to (7.1). So, since $R^{(d)}_{(n+1)i} := \tilde{R}^{(d)}_{Si} > \tilde{R}^{(d)}_{ki}$ for all $k \in S$ during the sybil attack and $a R^{\max} \leq 1 - a$ (by hypothesis),

$$
\tilde{R}^{(1)}_{(n+1)i} \leq R^{(1)}_{(n+1)i}.
$$

As in (7.3), for two different peers $j, i \neq n+1$ (i.e., $j, i \leq n$):

$$
\begin{aligned}
\tilde{R}^{(1)}_{ji} &:= a \sum_{k \notin S \cup \{i,j\}} R^{(d)}_{jk} R^{(d)}_{ki} + a \sum_{k \in S} \tilde{R}_{jS} \tilde{R}^{(d)}_{ki} + (1-a) R^{(d)}_{ji} \\
&= a \tilde{J}^{(1)}_{ji} + (1-a) R^{(d)}_{ji},
\end{aligned}
$$

and conclude as above.

□

The extent to which multi-step referrals are sybil-proof can be ascertained inductively using the arguments above.

7.3   DISCUSSION

Reputation frameworks are not immune to non-sybil false referrals (i.e., those *not* subject to (7.1)). Generally, false-referral problems can be limited by feedback of transaction outcomes to impact the direct reputations of associated *referring* peers (or their "referring" reputations used to weight their referrals), e.g., the anti-spam for voice over Internet telephony (anti-SPIT) referral framework of [120]. A basic assumption of such feedback systems is that the referrals are a factor in future transactions whose outcomes are, individually, not very important. See Ch. 10 for a consensus mechanism suitable for the case of important transaction outcomes.

CHAPTER 8

Peer-to-peer file sharing

In this chapter, we focus on distributed BitTorrent file-sharing which produces enormous traffic volume in the Internet [137]. BitTorrent presently does not employ referrals, although there is a type of Boolean reputation, based on the outcomes of bidirectional file-segment trading transactions, that forms the basis of a "choking" system locally within each BitTorrent swarm.

8.1   OVERVIEW OF BITTORRENT

To share files via BitTorrent, the initial file distributor creates a *.torrent* file which specifies the tracker and describes how the file is partitioned into small pieces, called blocks[1]. A block is a unit of data exchange in BitTorrent. The tracker is an online server that keeps track of the swarm of peers uploading and downloading the blocks of a particular file. The initial distributor posts the *.torrent* file on a file server accessible to peers that are interested in the file, e.g., on a web server or through a structured (DHT-based) peer-to-peer system[2].

To download via the BitTorrent protocol, a peer downloads the *.torrent* file associated with the file of interest and the peer's client then contacts the tracker to retrieve the list of peers sharing the file (peer set). Peers on this list are then contacted.

A peer maintains a list of number of copies of each block in its peer set. Assume that m is the number of copies of the rarest block. All blocks with m copies form a rarest block set. The peer randomly selects a block to download from the rarest block set, i.e., a *locally rarest-first* block-dissemination rule [22].

A peer may acquire a block through a unilateral (one-way client-server) transaction from a *seeder* (i.e., a peer who possesses all blocks and is actively engaged in the swarm), or by swapping blocks with a fellow *leecher* (i.e., a peer who needs blocks to complete his file). In BitTorrent, tit-for-tat choking rules are used to penalize peers that transmit their blocks at rates substantially lower than their partner peers in swap transactions [9, 36]. That is, peers essentially rank one another as a result of these transaction outcomes. A result of choking is that peers are eventually *clustered* into groups with similar uplinks [95, 127]. To turn such systems into *incentives* for cooperative behavior, peers with unacceptably (to them) poor transaction performance can be given the opportunity to *rehabilitate* their reputation by increasing the uplink bandwidth they allocate for file sharing; to this end, one peer can "optimistically" unchoke another (i.e., remove it from its choke list). Clearly,

---

[1]In BitTorrent, blocks are distributed along with a (hashed) signature for authentication purposes.
[2]Again, though keyword-based search is more difficult in the p2p framework, it can continue to function to find *.torrent* files even when the current web servers and associated search mechanisms are shut down by litigating copyright holders.

choking will not be an effective incentive mechanism if leechers can obtain the majority or all of the blocks of a file from the seeders in the swarm.

8.2 FILE SEGMENTATION

STORE-AND-FORWARD ECONOMIES

One motivation behind file segmentation[3] is that it reduces the store-and-forwarding costs of file dissemination. To see this for only client-server transactions, consider a group of n identical peers arranged in a highly idealized symmetric tree structure of depth $\log_k n > 1$. That is, each peer but those at the leaves are a tributary to $k < n$ others. Thus, there are a total of $1 + k + k^2 + ... + k^{\log_k n} \approx n$ peers. Suppose a file of length L is to be transmitted first by the peer located at the root.

The duration of time required, from the commencement of transmission by the root until complete reception by all other peers, is proportional to

$$kL \log_k n,$$

where we are accounting for store and forward delays at each level in the tree, i.e., the file must be completely received before it can be forwarded further down the tree toward the peers located at the leaves.

Now suppose that the file is segmented into m blocks each of length L/m. The total time required is now reduced, because of reduced store-and-forward delays of the smaller blocks, to

$$kL + k\frac{L}{m}(-1 + \log_k n).$$

That is, in a pipelined fashion, one block can be transmitted (again, after being wholly received) while the others are still being received or awaiting transmission from the tributary nodes. Thus, segmentation into blocks changes the total delay by a factor of[4]

$$\frac{kL + k(L/m)(-1 + \log_k n)}{kL \log_k n} = \frac{m - 1 + \log_k n}{m \log_k n} < 1.$$

If seeders share with the leechers a small volume of critical blocks (particularly the currently rarer ones) in a one-way server-to-client fashion, then the leechers can more rapidly disseminate these blocks primarily through trading/swapping transactions. In BitTorrent, the seeders may gain nothing by their actions. They are present because:

- they are often not using their work-stations and their Internet access is based on flat-rate charges for unlimited use (limits are only in terms of uplink and downlink capacity); and

- their liability for copyright infringement is potentially limited by:

---

[3] And message packetization, recall Sec .1.1.
[4] The inverse of which is also known as an ideal "multicasting gain".

- the file segmentation framework itself, i.e., disseminating only one small piece of a file at any given time; and

- a third-party swarm discovery system, e.g., downloading *.torrent* files via certain web sites.

FILE-SWAPPING IN A TWO-BLOCK SWARM

We now describe a deterministic model used in epidemiology (*cf.* Ch. 11) for a two-block swarm of completely connected users with comparable uplink bandwidth allocated for file sharing. Let $\lambda_\varnothing$ be the total arrival rate to a specific swarm of new leecher peers who possess no blocks ($\varnothing$) of the data object F being disseminated in the swarm. Let λ_F be the arrival rate of seeder peers who, again, possess the entire file F. We assume

$$0 \; < \lambda_F \; \ll \; \lambda_\varnothing.$$

Thus, we have identified two types of seeders:

- *temporary* seeders, i.e., those who arrived at the swarm as leechers and, upon downloading the complete file and becoming seeders, will quickly depart[5]; and

- *permanent* seeders who, in our model, are a smaller population of peers who continually depart and return as seeders.

Let $1/\delta_F$ be the mean sojourn time of permanent seeders, giving an average population λ_F/δ_F of permanent seeders by Little's theorem [157]. Similarly, define $1/\delta_\varnothing$ as the mean sojourn time (as seeders) of temporary seeders, with $1/\delta_\varnothing \ll 1/\delta_F$. The mean lifetime $1/\delta$ of a typical seeder is therefore an average weighted by the mean population sizes[6],

$$\frac{1}{\delta} \; = \; \frac{\lambda_\varnothing/\delta_\varnothing}{\lambda_\varnothing/\delta_\varnothing + \lambda_F/\delta_F} \cdot \frac{1}{\delta_\varnothing} + \frac{\lambda_F/\delta_F}{\lambda_\varnothing/\delta_\varnothing + \lambda_F/\delta_F} \cdot \frac{1}{\delta_F} \; .$$

Let β be a rate parameter of client-server transactions, which depends on the size of the file being transmitted and the associated willingness of the server peer to participate in the transaction (for nothing in return from the client peer). Let γ be a rate parameter of swap transactions. We assume here that the swaps occur more readily than client-server transactions, i.e.,

$$\gamma \; \geq \; \beta. \tag{8.1}$$

The data transmission rates (uplink bandwidths) employed by different peers (for block-uploading transactions) are not directly considered in the following model (see the discussion of intra-swarm clustering below). The successful transaction rate is here modeled as proportional to the contact rate between the peer populations that can engage in the transaction. In turn, the contact rate between

---

[5]Once the human operator returns to their work-station [21].
[6]Each term λ/δ is by Little's theorem.

a pair of populations is simply assumed to be proportional to the product of the population sizes [41, 48], where β and γ may be decreasing in the total number of peers in the swarm. Other applicable and similar families of models have also been extensively studied, e.g., urn, replicator, Volterra-Lotka, and coupon-collector [102].

Suppose F is split-up into two blocks indexed 1 and 2. Let

$$\underline{x} = (x_\varnothing, x_1, x_2, x_F) \in [0, \infty)^4$$

be the state of the system. In particular, the seeder population at time t is approximately the non-negative real-valued quantity $x_F(t)$. So, the model for this system is:

$$\dot{x}_\varnothing = \lambda_\varnothing - \beta x_\varnothing(x_1 + x_2 + x_F)$$
$$\dot{x}_1 = -x_1(\beta x_F + \gamma x_2) + \beta x_\varnothing(x_1 + \tfrac{1}{2}x_F)$$
$$\dot{x}_2 = -x_2(\beta x_F + \gamma x_1) + \beta x_\varnothing(x_2 + \tfrac{1}{2}x_F)$$
$$\dot{x}_F = \lambda_F + \beta(x_1 + x_2)x_F + 2\gamma x_1 x_2 - \delta x_F.$$

Here, all client-server transactions involve the β factor and the swapping transaction involves the γ factor. Also note that half of the successful transactions between $x_\varnothing$ and x_F result in an arrival to the x_1 population (that possess only the first block of F), and the other half result in an arrival to the x_2 population. This equity assumption is justified by the locally rarest-first dissemination rule (so that all blocks tend to be equally rare among the leechers in the swarm) and is further discussed below.

Claim 8.1 The unique and stable equilibrium of the two-block case is

$$\underline{x}^* := (x_\varnothing^*, x_1^*, x_2^*, x_F^*)$$
$$= \left(\frac{\lambda_\varnothing}{\beta} \left(\sigma + \frac{\lambda_\varnothing + \lambda_F}{\delta} \right)^{-1}, \frac{\sigma}{2}, \frac{\sigma}{2}, \frac{\lambda_\varnothing + \lambda_F}{\delta} \right),$$

where

$$\sigma = \frac{\beta(\lambda_\varnothing + \lambda_F)}{\gamma \delta} + \sqrt{\left(\frac{\beta(\lambda_\varnothing + \lambda_F)}{\gamma \delta} \right)^2 + 2\frac{\lambda_\varnothing}{\gamma}}.$$

Proof. Solve $\underline{\dot{x}} = 0$ (i.e., the above dynamics at equilibrium) by first noting that adding all four equations at equilibrium gives

$$0 = \lambda_\varnothing + \lambda_F - \delta x_F^* \quad \Rightarrow \quad x_F^* = \frac{\lambda_\varnothing + \lambda_F}{\delta},$$

which is consistent with Little's theorem. Second, note by symmetry that

$$x_1^* = x_2^* =: \sigma/2.$$

So, $\dot{x}_\varnothing = 0$ implies

$$x_\varnothing^* = \frac{\lambda_\varnothing}{\tilde{\beta}}\left(\sigma + \frac{\lambda_\varnothing}{\delta}(1 + \lambda_F/\lambda_\varnothing)\right)^{-1}.$$

Substitute the last two displays into, say, $\dot{x}_1 = 0$ to get that σ is the unique positive solution of

$$0 = \sigma^2 + \frac{2\beta\lambda_\varnothing}{\gamma\delta}(1 + \lambda_F/\lambda_\varnothing)\sigma - 2\frac{\lambda_\varnothing}{\gamma}.$$

Finally, stability can be checked by applying the Hartman-Grobman theorem.

$\square$

Using Little's theorem, we can use Claim 8.1 to find a closed-form expression for the sojourn time from leecher to seeder (i.e., the time to acquire both blocks), $(x_\varnothing^* + x_1^* + x_2^*)/\lambda_\varnothing$. Using this result, parametric conditions can be given for when it is advantageous (in terms of minimizing mean complete file acquisition times) to split a file into blocks, compared to simple client-server interactions with the available seeders [88].

For a stochastic transaction-level swarm with more than two blocks, by scaling time and space one can show how such a deterministic model above is its "fluid" limit; see [88] and Appendix A of [48].

We can extend the model to a collection of $K > 1$ different sub-groups of a swarm, each modeled as above and each corresponding to a range of allocated uplink bandwidth [95, 127]. Within each sub-group, all peers can directly transact with each other and the locally rarest first block-dissemination policy should balance segment instances in the sub-group (consistent with $x_1^* = x_2^*$ above). These sub-groups could be modeled as sharing the permanent (π) seeder population obeying

$$\dot{x}_F^{(\pi)} = \lambda_F - \delta^{(\pi)}x_F^{(\pi)},$$

with equilibrium population $x_F^* = \lambda_F/\delta_F$. The kth sub-group's model would, for example, have temporary seeder dynamics,

$$\dot{x}_F^{(k)} = (x_1^{(k)} + x_2^{(k)})(\beta^{(k)}x_F^{(k)} + \beta^{(\pi)}x_F^{(\pi)}) + 2\gamma^{(k)}x_1^{(k)}x_2^{(k)} - \delta^{(k)}x_F^{(k)},$$

for $k \in \{1, 2, ..., K\}$.

Computation of the equilibrium of each swarm (or swarm sub-group k) with arbitrary $n \geq 2$ blocks is simplified to solving for just $n + 1 \ll 2^n$ quantities x_A^*, where A is an index subset of the n blocks, because $x_A^* = x_B^*$ whenever $|A| = |B|$ under rarest-first block-dissemination. These models under slowly time-varying leecher arrival rates $\lambda_\varnothing^{(k)}$ could be evaluated at equilibrium expressed as functions of these parameters, i.e., the system observed in quasi-stationarity.

DISCUSSION OF FILE SEGMENTATION AND DISSEMINATION

Given seeder departures, the presence of permanent seeders was modeled above by the assumption that $\lambda_F > 0$, i.e., a small but persistent arrival rate of permanent seeders returning to the swarm. A large number of uncongested seeders may undermine the types of cooperative behavior among leechers that are fostered by choking rules and segmentation to promote swapping behavior. Rarest-first block dissemination is intended to reduce the probability of block extinction (moreover, to achieve equally common segments) and to equip leechers with valuable blocks to swap (note, however, that the presence of seeders will prevent the extinction of any block [21]).

So, rather than rarest-first block dissemination, a policy that deliberately creates rare blocks will naturally *increase* acquisition times of the whole file by leecher peers who are disinclined to remain in the swarm once they become seeders. To model this in the two-block swarm model above, the factors of $\frac{1}{2}$ can be replaced by α and $1 - \alpha$ in the expressions for $\dot{x}_1$ and $\dot{x}_2$, respectively [65]. If $\frac{1}{2} < \alpha \leq 1$, then the dissemination of the first block will be favored and the second block will be rarer, i.e., $x_2^* < x_1^*$.

8.3   DISCUSSION

BitTorrent's choke lists correspond to intra-swarm reputations. One can conceive disseminating choke lists in the social graph of participating BitTorrent users thereby creating a referral system which can also be used inter-swarm. Such referrals may trade-off rehabilitation opportunities (optimistic unchoking) with quicker reaction to leechers with low allocated uplink bandwidths. When choking-related incentives do not work because many leecher peers are simply unwilling or unable to increase their uplinks and when the seeders are congested, unchoking may allow peers to access needed blocks even if it means acquiring them at a rate significantly slower than their own allocated uplinks for file-sharing.

The *Vuze* BitTorrent client employs the *Vivaldi* coordinate system wherein a certain small core subset of well spaced peers act as landmarks. The remaining peers of a swarm use the round-trip-times of pings to ascertain the most physically proximal landmark. The peers in a swarm are thus clustered by common landmark and will tend to trade blocks with each other. Such trading is more efficient because of the physical proximity of peers within a cluster.

In a file-sharing system that employs "network coding", a peer combines all the blocks they possess into a single unit for purposes of dissemination. So, when a peer with block subset A (counted in x_A) has a single transaction with a peer with block subset B, the result is that the two peers now possess $A \cup B$ (i.e., two peers arrive to the community counted $x_{A \cup B}$).

Finally, we mention the method of threshold cryptography wherein a secret key is partitioned into k segments and the segments are distributed among $n > k$ peers (one segment per peer). So, a cooperating group of peers has to be sufficiently large (i.e., sufficiently "trustworthy") to be able to reconstruct the key. For the case of purely random dissemination of key-segments, the probability that a randomly chosen set of m different peers are able to reconstruct the key is

$$p(m) \quad := \quad \binom{(m-k)+k-1}{k-1} \bigg/ \binom{m+k-1}{k-1} = \binom{m-1}{k-1} \bigg/ \binom{m+k-1}{k-1},$$

where $\binom{m+k-1}{k-1}$ is the number of different combinations of k different objects after m choices with replacement. So, the average number of peers required to reconstruct the key is

$$\mathsf{E}M \quad := \quad \sum_{m=k}^{n} m p(m) \bigg/ \sum_{m'=k}^{n} p(m').$$

In one approach, fix n and select k so that a certain number of peers (say a majority on average, i.e., $\mathsf{E}M > n/2$) is required to reconstruct the key. Note that the probability that not all key segments are distributed among the n peers (something that would not happen in practice) is $1 - p(n)$, but $1 - p(n) \approx 0$ for $n \gg k$. Threshold cryptography can be implemented as a kind of coupon collecting system at a group level.

PART IV

Consensus

CHAPTER 9

Consensus in dynamical belief systems

A referral or recommender system, as discussed in Ch. 6, can be interpreted as trying to create a consensus about the reputation of an individual beyond the individual's immediate neighbors (social community). The model that tracks the stage of a disease or the fraction of a data object (or number of coupons) currently possessed by an individual can be reinterpreted to model the degree to which an individual agrees with an issue. There is a large literature on models that focus on gossip propagation, e.g., [26, 134], and related gossip/epidemic protocols for information dissemination. Following Sec. 6.2, in this chapter we focus on the development of a global consensus regarding a single issue. Significantly more complex information exchange models than what we describe in the following have been considered in a social networking context, including information gleaned from passive observations, e.g., see Sec. 2 of [2].

9.1 NEAREST NEIGHBOR AVERAGING

Again, a vertex here will typically represent an individual peer, or possibly a "tight" social group, and an edge will represent a social tie. Social ties are generally neither symmetrical nor transitive, and this asymmetry is intuitively fundamental to consensus in social networks: peer i's opinion may be shaped by an influential person j (as may be indicated by $w_{i,j} = 1$), but j may not value i's opinion at all ($w_{j,i} = 0$). Note that we have reversed the order of the edge weight w's subscripts here. Let d_i be the number of different peers whose opinion i values, i.e., its out-degree $d_i := \sum_{j \neq i} w_{i,j}$. We'll take $w_{i,i} \equiv 1$ for the consensus context below.

Suppose that the social players act to exchange their current opinion about a particular issue, expressed as a real-valued quantity, θ. For example, $\theta_i = 0$ means that peer i is completely negative on the issue, whereas $\theta_i = 1$ means that i is completely positive. Further, suppose that the opinion held after the kth such iteration by peer i is

$$\theta_i(k+1) = \alpha\theta_i(k) + (1-\alpha) \sum_{j \neq i} \theta_j(k) \frac{w_{i,j}}{d_i}, \tag{9.1}$$

where $0 < \alpha < 1$ is a common parameter. So, we simply assume here that each peer *equally* weights the current opinions of his/her neighbors and the result is weighted $(1-\alpha)$ to α with his/her own current opinion. Again, recall our discussion of similar referral dynamics in Sec. 6.2, i.e., the "particular issue" in question was the reputation of a particular vertex (peer), where the opinions were weighted using the different "direct" reputation values of the different neighboring peers. Also note that we have arranged so that opinions are averaged (9.1). Weights α and w can be more freely chosen if we take a function of the right-hand side of (9.1) whose range is $[0, 1]$, e.g., $\frac{2}{\pi}\arctan(\cdot)$, before equating to $\theta_i(k + 1)$; the result is a Hopfield neural network operating in discrete time [70].

9.2 CONVERGENCE OF CONSENSUS

For a simple proof of convergence of consensus belief vectors $\underline{\theta}$, in addition to synchronous peer communication also assume that the social graph is connected, i.e., there is a (directed) path between any pair of peers i, j. Here, a directed edge *from i to j* is present if and only if $w_{i,j} > 0$ (i.e., the edge is in the opposite direction from the flow of opinion). For n peers, define the $n \times n$ matrix $\mathbf{P}$ with entries

$$P_{i,j} = \begin{cases} \alpha & \text{if } j = i \\ (1 - \alpha)w_{i,j}/d_i & \text{if } j \neq i. \end{cases}$$

In vector form,

$$\underline{\theta}(k + 1) = \mathbf{P}\underline{\theta}(k) = \mathbf{P}^{k+1}\underline{\theta}(0),$$

where $\underline{\theta}(0)$ is the vector of initial opinions.

Note that $\mathbf{P}$ is *row stochastic* and recall the irreducible and aperiodic conditions of a Markov chain[1] regarding convergence to its invariant distribution $\underline{\pi}$. Here, convergence is to the final opinions

$$\theta_i(\infty) = \sum_j \pi_j \theta_j(0)$$

for all peers i, i.e., a consensus of the inner product of $\underline{\pi}$ and $\underline{\theta}(0)$. In vector form,

$$\underline{\theta}(\infty) = \lim_{k \to \infty} \mathbf{P}^k \underline{\theta}(0) = \mathbf{P}^\infty \underline{\theta}(0).$$

For example, if $\alpha = 0$ and if $\mathbf{P}$ is nevertheless aperiodic, then there is a unique invariant distribution: $\pi_i = d_i/D$ where $D := \sum_j d_j$. So, the final consensus opinions are

$$\sum_i \theta_i(0)\frac{d_i}{D}.$$

The consensus convergence result above does not require equally weighting the opinions of neighboring peers as in the example $\mathbf{P}$ above and, again, other types of averaging are possible, e.g., [26]. The discussion of Sec. 5.1 is relevant to the question of how rapidly this convergence may occur if we take $\alpha = 0$ and assume social ties are symmetrical, i.e., $w_{i,j} = w_{j,i}$ for all i, j.

---

[1]Which are met as a consequence of the assumptions in Sec. 5.1.

If the social graph is not connected, then $\mathbf{P}$ is the aperiodic transition probability matrix of a Markov chain whose state-space (the peers or vertices) can be partitioned into multiple disjoint "communicating classes" and the transients [67]. Suppose some peers i do not change their initial opinions $\theta_i(0)$, while others act as above, so that the unchanging-opinion peers are basically *inputs* to the consensus system. These unchanging-opinion peers correspond to absorbing states of the Markov chain, i.e., singleton communicating classes. The opinions of communicating classes will influence those of the "transient" peers in steady-state. For example, suppose there are five peers with

$$\mathbf{P} = \begin{bmatrix} 1 & 0 & 0 & 0 & 0 \\ \frac{1}{6} & \frac{1}{2} & \frac{1}{6} & \frac{1}{6} & 0 \\ 0 & \frac{1}{4} & \frac{1}{2} & \frac{1}{4} & 0 \\ 0 & \frac{1}{6} & \frac{1}{6} & \frac{1}{2} & \frac{1}{6} \\ 0 & 0 & 0 & 0 & 1 \end{bmatrix}.$$

Here, peers indexed 1 and 5 have unchanging opinions and the rest are transients with $\alpha = 1/2$. Thus[2],

$$\mathbf{P}^\infty := \lim_{n \to \infty} \mathbf{P}^n = \begin{bmatrix} 1 & 0 & 0 & 0 & 0 \\ \frac{5}{8} & 0 & 0 & 0 & \frac{3}{8} \\ \frac{1}{2} & 0 & 0 & 0 & \frac{1}{2} \\ \frac{3}{8} & 0 & 0 & 0 & \frac{5}{8} \\ 0 & 0 & 0 & 0 & 1 \end{bmatrix}.$$

So, peer 2 will weight peer 1's opinion by the factor $\frac{5}{8}$ and peer 5's opinion by $\frac{3}{8}$, peer 4 will weight peer 1's opinion by the factor $\frac{3}{8}$ and peer 5's opinion by $\frac{5}{8}$, while peer 3 will equally weight (simply average) the opinions of peers 1 and 5.

9.3   GIBBS INVARIANT OF A MARKOV RANDOM FIELD

Markov random fields have also been used to model social consensus [53, 66, 85]. Consistent with marginals describing local influence (spatial consensus correlation, including the conditional random fields framework) and conditional independence (spatial Markov property), a global joint consensus distribution is directly sought without considering the limiting behavior of a time-varying Markov chain. Communicating classes (cliques) are handled through the Hammersley-Clifford theorem. "Bayesian networks" and "belief propagation" are similar to Markov random fields and have also been applied to ranking in peer-to-peer social networks, e.g., [20].

---

[2]Note that $\mathbf{PP}^\infty = \mathbf{P}^\infty \mathbf{P} = \mathbf{P}^\infty$. Also, for transition probability matrix $\mathbf{P}$ and initial state 2, a Markov chain will absorb in state 1 with probability $\frac{5}{8}$, otherwise absorb at state 5 with probability $\frac{3}{8}$; see the forward conditioning argument in Sec. 4.7 of [79].

For a simple example, let the current Boolean opinion state of vertex i be $x_i \in \{-1, 1\}$ and assume symmetric and Boolean edge weights on the connected graph: $w_{i,j} \in \{0, 1\}$ with $w_{i,i} = 0$ for all i, i.e., no self-loops. Define the "energy" on the graph as $U(\underline{x})$ and the associated Gibbs probability distribution on collective opinions $\underline{x}$,

$$\Gamma(\underline{x}) \quad = \quad \frac{e^{-U(\underline{x})/T}}{Z}, \tag{9.2}$$

where the normalization term (partition function) $Z = \sum_{\underline{x}} e^{-U(\underline{x})/T}$. Since the (temperature) parameter $T > 0$, the Gibbs distribution will favor opinion vectors $\underline{x}$ with smaller energy U.

The entropy of $H(p) = -\sum_{\underline{x}} p(\underline{x}) \log p(\underline{x})$ of a distribution p is maximized by the uniform or "unbiased" distribution, $p(\underline{x}) = 2^{-n}$ for all $\underline{x} \in \{-1, 1\}^n$. The Gibbs distribution Γ is the "least biased" distribution in the sense that it achieves the minimum entropy among those distributions with the same mean energy $EU(\underline{X})$, where $\underline{X}$ is the collective-opinion random variable, see Ch. III of [85] and Ch. 7 of [66]. Also, the Gibbs distribution possesses the following spatially Markovian property: for all peers i,

$$\mathsf{P}(X_i = \theta_i | X_j \text{ for all } j \neq i) \quad = \quad \mathsf{P}(X_i = \theta_i \mid X_j \text{ such that } w_{i,j} = 1). \tag{9.3}$$

That is, given the opinions of the whole group, the opinion of any peer i will depend only on those of its neighbors.

For a simple one-dimensional Ising model example, suppose that $n + 1$ peers are arranged in a row and that the left-most peer has opinion $\in \{-1, 1\}$ with equal probability (a kind of assumed "prior" probability). Further suppose that independently of other peers, a peer agrees with the neighbor on their left with probability $p > 0.5$ (i.e., the two will tend to agree), otherwise disagrees with probability $1 - p$. Define the energy U so that the total number of adjacent disagreements in the collective opinion vector $\underline{x}$ is $(n + U(\underline{x}))/2$. Equivalently, $U(\underline{x}) = -\sum_{i<j} x_i x_j w_{i,j}$ which is larger when there is more disagreement among adjacent peers (i.e., when $x_i x_j w_{i,j} < 0$). We have by independence that the probability of a certain collective opinion $\underline{x}$ of $n + 1$ peers is

$$\frac{1}{2}(1 - p)^{(n+U(\underline{x}))/2} p^{(n-U(\underline{x}))/2} \quad = \quad \frac{1}{Z} e^{-U(\underline{x})/T},$$

where $Z = 2((1 - p)p)^{-n/2}$ and $T = 2/\log(p/(1 - p))$ (note $T > 0$ because $p > 0.5$). That is, the distribution is Gibbs; see Chapter VI of [85].

There is a large literature on biasing techniques similar to those used in Sec. 5.2 but involving Gibbs invariant distributions, e.g., [69, 147].

CHAPTER 10

Byzantine consensus

In this chapter, we consider the problem of consensus of Boolean opinions regarding an *important* issue in the presence of faulty or malicious peers. Assume that there is a single peer indexed $n + 1$ who purports direct knowledge about some other peer indexed $x > n + 1$, where peer $n + 1$ is the "general" or "source" in the following Byzantine agreement problem setting. Suppose that peer $n + 1$ communicates some information v_k about x to peer k where $k \in \mathcal{N} := \{1, 2, ..., n\}$, i.e., n different "lieutenant" peers of the general peer, for the purposes of dealing with peer x. Further suppose that the lieutenants are meant to collectively act on this information about x and that action is in and of itself far more important than the peer-to-peer transactions previously considered here.

Two types of problems may arise in this setting. The first is that the general may communicate different values v_k about peer x to the n lieutenant peers. The second is that the lieutenant peers may not obey the following protocol intended to achieve their consensus. Peers that do not follow protocol, either deliberately through dishonest means or because they are faulty, are termed "incorrect". So, the goals of the algorithm discussed in this chapter are that, despite the presence of incorrect peers:

- the correct lieutenant peers reach consensus on the information about peer x which they received from the general peer, i.e., they agree on the same value v^*; and

- if the general peer has communicated the same value $v_k = v$ at least to all correct peers (i.e., the general peer itself is correct in this sense), then $v^* = v$.

The following algorithm requires identity authentication mechanisms, reliable point-to-point communications, and synchronization among the n lieutenant peers. The goal of Boolean consensus can be met if

$$n \geq 3w + 1,$$

where w is the number of incorrect peers.

10.1 THE TREE DATA STRUCTURE

In the approach of [16, 93], a tree is constructed by each of the n lieutenant peers. Every different direct path from root to leaf in the tree will consist of $w + 1$ nodes (vertices): the root and w nodes chosen without replacement from the population of $n - 1$ other lieutenants. That is, the tree will have $(n - 1)!/(n - 1 - w)!$ paths (and hence that many different leaves), and the index k of the lieutenant node that is constructing the tree is excluded (hence '$n - 1$' is used instead of 'n').

For lieutenant peer k's tree, every interior node at *depth d* from the root (where $1 \leq d \leq w - 1$) will correspond to a permutation of d nodes chosen from the $n - 1$ other lieutenants, and so is labeled as the ordered $(d + 1)$-tuple $\underline{v} := (v_0, v_1, ..., v_{d-1}, v_d)$, where:

- the root $v_0 \equiv n + 1$;

- $\underline{v}$ corresponds to the direct path from root v_0 and for all $i > 0$ the $v_i \in \mathcal{N} \setminus \{k\}$ are all different;

- the parent node at depth $d - 1$ is labeled $(v_0, v_1, ..., v_{d-1})$; and

- the $n - d - 1$ children nodes are labeled $(v_0, v_1, ..., v_d, q)$ for $q \in \mathcal{N} \setminus (\{k\} \cup \underline{v})$, i.e., one child node for each lieutenant peer q *not* in $\underline{v}$.

The leaf nodes at depth w will obviously have no children.

Each node/vertex v of lieutenant k's tree is associated with two values: $input_k(\underline{v})$ and $output_k(\underline{v})$.

10.2 INFORMATION GATHERING TO DETERMINE $input(v)$

In round zero of the information gathering phase, the general peer $n + 1$ communicates the values v_k to the lieutenant peers k. Again, the root of all trees is labeled $n + 1 \equiv v_0$, and each peer k sets $input_k(n + 1) = v_k$.

In subsequent, synchronized communication rounds indexed $d > 0$, the n lieutenant peers will exchange information stored at depth $d - 1$ in their trees. For all correct peers k and v_d,

$$input_k(v_0, v_1, v_2, ..., v_{d-1}, v_d) = input_{v_d}(v_0, v_1, v_2, ..., v_{d-1}),$$

where, by design, $k \neq v_d$. That is, the *input* information stored at peer v_d at its node $\underline{v} := (v_0, v_1, v_2, ..., v_{d-1})$ is communicated to *all* other peers $k \in \mathcal{N} \setminus (\{v_d\} \cup \underline{v})$ and, upon receipt, those peers store this information at node $(v_0, ..., v_{d-1}, v_d)$ in their own tree. Ostensibly, $input(\underline{v})$ is simply what lieutenant peer v_{d-1} says v_{d-2} says ... v_2 says v_1 says the general peer $(n + 1 = v_0)$ said.

So, in round $0 < d < w$ of the information exchange phase, each lieutenant peer k communicates a potentially very large amount of information proportional to $(n - 1)!/(n - d - 1)!$, i.e., the number of nodes at depth d in its tree, *cf.* Sec. 10.5.

10.3 LEAVES-TO-ROOT MAJORITY VOTING

Consider a discrete range of possible information values $V = \{0, 1, \tilde{v}\}$, i.e., $v_k \in V$, where, to deal with incorrect peers, the special value $\tilde{v}$ is used to denote the absence of a value or one out of the nominal Boolean $\{0, 1\}$ range.

In this phase, consensus is built among the correct lieutenant peers through leaf-to-root majority voting. Let $C_k(\underline{v})$ be the set of $n - 1 - d$ children nodes of the non-leaf node $\underline{v} := (v_0, v_1, ..., v_{d-1}, v_d)$ of peer k's tree, i.e., $C_k(\underline{v}) = \{(\underline{v}, q) \mid q \notin \{k\} \cup \underline{v}\}$. Let $M(S)$ be the majority of values of the finite set S, if a majority exists.

Each peer k separately processes their received $input_k$ values stored in their tree's nodes to determine the $output_k$ vertex values as follows:

$$output_k(\underline{v}) \ = \ \begin{cases} input_k(\underline{v}) & \text{if } \underline{v} \text{ is a leaf} \\ M(\{output_k(z) \mid z \in C_k(\underline{v})\}) =: v & \text{if the majority } v \text{ exists} \\ & \text{and } v \neq \tilde{v} \\ \tilde{v} & \text{else.} \end{cases}$$

10.4 FUNDAMENTAL CONSENSUS PERFORMANCE RESULT

The notion of agreement/consensus pertains to the computed output values at the roots, $output_k(n + 1)$, for all peers $k \in \{1, ..., n\}$. Recall that w is the number of incorrect lieutenant peers. Let $|\underline{v}| \geq 0$ be the depth of (number of lieutenant peers indexing) node $\underline{v}$.

Lemma 10.1   *If $n > 3w + 1$ and $\underline{v}$ is an internal node, then $\underline{v}$ has $n - |\underline{v}| - 1 \geq 2w + 1$ children.*

Proof. This is simply due to the fact that the depth of the trees constructed is $\leq w$, so that $|\underline{v}| \leq w$.
□

So, clearly if all but w children of an (internal) node $\underline{v}$ have common output values, then that common value will be the majority among the children of $\underline{v}$ and hence it will become the output value $\underline{v}$. Thus, tolerance of fewer than $(n - 1)/3$ incorrect peers is argued by reverse induction. A node $\underline{v}$ is said to have *consensus* if $output_k\underline{v}$ is the same for all correct peers $k \notin \underline{v}$ (i.e., all correct peers that possess node $\underline{v}$ in their tree).

Claim 10.2   If $n > 3w + 1$, then:

- the root nodes of the correct lieutenant peers k will have consensus, say with value $v^* = output_k(n + 1)$; and

- if the general is correct, so that all correct lieutenant peers k have a common input root value $input_k(n + 1)$, then $input_k(n + 1) = v^*$ for all correct lieutenants k.

Proof. First consider any correct lieutenant peer q and any *leaf* node $(\underline{v}, q)$, i.e., $|\underline{v}| = w$. Since q is correct, $input_q(\underline{v}) = input_k(\underline{v}, q)$ for all $n - w - 1$ peers $k \notin (\underline{v}, q)$. For all such peers k that are correct,

$$output_k(\underline{v}, q) \ = \ input_q(\underline{v}) \ = \ input_k(\underline{v}, q). \tag{10.1}$$

That is, if q is correct, then there is consensus among all leaf nodes $(\underline{v}, q)$ of correct peers $k \notin (\underline{v}, q)$.

Now for arbitrary $d < w$, inductively assume that all nodes $(\underline{v}, q)$ such that $|\underline{v}| = d$ (i.e., $\underline{v} = (v_0, v_1, ..., v_d)$ for d different peer nodes v_i) have consensus whenever q is correct.

To complete the (backward) inductive portion of the proof of the first statement of the claim, we need to show that the correct parent nodes of $\underline{v}$ have consensus whenever v_d is correct. By the previous lemma, the number of children of the parent node $(v_0, v_1, ..., v_{d-1})$ is $\geq 2w + 1$ where w is the number of incorrect peers. Thus, the majority terminate with the index of a correct peer. By this inductive argument, the root nodes (at depth $d = 0$) have consensus.

For the second statement of the claim, repeat the inductive argument further assuming that for all $|\underline{v}| = d < w$ and all correct peers q:

$$output_k(\underline{v}, q) = input_k(\underline{v}, q)$$

for all correct peers $k \notin (\underline{v}, q)$ (which holds for leaf nodes by (10.1)). So, if lieutenant peer v_d is correct, then since k, q are also correct:

$$input_k(\underline{v}, q) = input_q(\underline{v}) = input_{v_d}(\underline{v}\backslash v_d) = input_k(\underline{v})$$
$$\Rightarrow output_k(\underline{v}, q) = = input_k(\underline{v}).$$

Now, since the majority of node $\underline{v}$'s children in k's tree have a correct terminating peer q, and *all* such children will therefore have **output** values equal to $input_k(\underline{v})$, the majority rule gives $output_k(\underline{v}) = input_k(\underline{v})$.

So by induction, if the general peer $n + 1$ is correct, then (for $d = 0$)

$$output_k(n + 1) = input_k(n + 1)$$

for all correct peers k.

□

10.5 DISCUSSION

The following generalization is a step toward a consensus algorithm with overhead of polynomial complexity; see Lemma 5.3 of [62] (also, e.g., Sec. 5.2.5 of [13]). Recall that round d requires $O((n - 1)!/(n - d)!)$ communication per peer. Overall communication overhead in round d is therefore exponential in d. Clearly, therefore, communication overhead is also $O(n^d)$ in the number of lieutenant peers, n. For a node $\underline{v}$ at depth d, suppose that if $F(d)$ of its children have common **output** value then this value becomes the **output** value of $\underline{v}$, where

$$w + 1 \leq F(d) \leq n - w - d.$$

The fundamental claim can be directly extended to **output** functions using such an F. Note how the simple majority function used above,

$$M(d) = \left\lfloor \frac{n - d}{2} \right\rfloor + 1,$$

is an example of such a function F. A judicious choice of F will give *early prediction* of the majority on a path, thereby reducing the overhead by making unnecessary any further communication along that path.

The problems of asynchronous consensus in the presence of faulty participants have been extensively explored, including robust gossip by reliable shared observation of a random variable; see [11] and Ch. 12-14 of [13]. Finally, note that if the lieutenant peers differently weight each other's data, we may arrive at a system similar to that discussed in Ch. 7.

CHAPTER 11

Epidemics

Epidemics of malicious software spread over the Internet are a serious threat to privacy, identity, and property, including computing and financial resources. Many different types of epidemiological models have been proposed, e.g., [41, 48]. Some of these models have also been used for file-sharing, consensus, or more generic frameworks for dissemination of information, e.g., [56]. In this chapter, we describe the online spread of viruses and worms using the type of model also used in Ch. 8 for file-sharing.

11.1 MODELS OF DISEASE SPREAD IN LIVING POPULATIONS

Disease spread in living populations depends on:

- the dynamics (including social aspects) of physical contact by infected individuals required to pass the disease, in turn depending on:

  - how the disease was introduced into the population; and

  - how the disease naturally affects the mobility of those infected, including the removal of infected individuals by death, in addition to pre-existing mobility factors such as geographical barriers;

- any incubation/dormancy time the disease requires before it is contagious i.e., an individual can be infected but not yet infective; and

- the probability of disease transfer given the necessary physical contact with an infective.

A particularly lethal disease may quickly destroy the local population into which it was introduced before an opportunity to spread outside that population presents itself.

Countermeasures to the spread of disease can target any of the factors just mentioned. For infected individuals, this could include:

- medical treatments to cure the disease or reduce its contagiousness; and

- physical interventions: from wearing of face masks to reduce contagiousness; to limitations to mobility, e.g., avoiding points of congregation such as bus terminals; to complete quarantine.

Also, uninfected individuals may take precautions, e.g., inoculation to reduce or eliminate suscep- tibility. Often countermeasures are delayed, e.g., by the need to test an inoculation for side effects.

Finally, diseases may mutate to circumvent medical treatments and acquired immunities, thus resulting in dynamics where some individuals continually transition among infective, recovered and susceptible states.

Some epidemic models simply attempt to capture how *average* populations of susceptibles and infectives evolve depending on certain important parameters. Stochastic approximations such as those given by Kurtz's theorem [92] can be used to relate simpler deterministic models to scaled limits of more detailed stochastic ones modeling infections of individuals; see [88] and Appendix A of [48].

SIR MODELS FOR A CLOSED POPULATION

Consider now a closed (fixed) population of N (susceptible) peers to a disease. At time t, let $x_i(t)$ be the number of peers infected with the disease, $x_s(t)$ be the number of uninfected peers, and $x_r(t)$ be the number of "removed" peers (dead/offline or cured/patched). Assume $x_s(0), x_i(0) > 0$ and $x_r(0) = 0$, and note that for all time $t \geq 0$,

$$x_s(t) + x_i(t) + x_r(t) \quad = \quad N. \tag{11.1}$$

The time-evolutions of these quantities can be continuously approximated using susceptibles-infectives-removals (SIR) dynamics [41, 48]: for times $t \geq 0$,

$$
\begin{aligned}
\dot{x}_s(t) &= -\beta x_s(t) x_i(t) \\
\dot{x}_i(t) &= \beta x_s(t) x_i(t) - \delta x_i(t) \\
&= [x_s(t) - \frac{\delta}{\beta}]\beta x_i(t) \\
\dot{x}_r(t) &= \delta x_i(t)
\end{aligned}
$$

with fixed parameters $\beta, \delta > 0$. Again, the product term $x_s x_i$ models how the rate of contact between the infective and susceptible populations is an increasing function of the size of each. Thus, the β parameter would be larger for higher probability of infection given contact. Also, the removal parameter δ represents the inverse of the lifetime of a peer as an infective. Note that summing the equations gives

$$\dot{x}_s(t) + \dot{x}_i(t) + \dot{x}_r(t) \quad = \quad 0 \text{ for all } t,$$

consistent with (11.1), and the equilibrium state $x_i(\infty) := \lim_{t \to \infty} x_i(t) = 0$. Finally, note that the above SIR model is clearly intended for a homogeneous population, i.e., all nodes equally susceptible and proximal to each other for purposes of disease spread, and with common disease lifetime.

The number of infectives will initially grow (i.e., $\dot{x}_i > 0$) if $x_s(0) > \delta/\beta$. The limiting values of x_r and x_s depend on the parameters $x_s(0)/N$ and the "relative" removal rate, δ/β [41, 48], particularly with regard to the question: "Will the disease infect all susceptibles, i.e., will $x_r(\infty) = N$ and $x_s(\infty) = 0$?" Combining the first and third equations gives

$$\dot{x}_s/x_s \quad = \quad -\dot{x}_r \delta/\beta,$$

so that

$$x_s(t) \quad = \quad x_s(0)e^{-x_r(t)\beta/\delta}.$$

Thus, positive $x_r(\infty) < N$ can be numerically solved from

$$N - x_r(\infty) \quad = \quad x_s(\infty) \quad = \quad x_s(0)e^{-x_r(\infty)\beta/\delta}.$$

So, we can assume that the infection reached the whole population only if real-valued x_r exceeds $N - 1$.

The previously described SIR model has only a few important parameters, and thus it's unlikely to "overfit" a given set of epidemiological data. Overfitting bias is a significant problem of more complex versions of these models [8, 110].

SIR MODELS FOR AN OPEN POPULATION

For exogenous arrival of susceptibles[1] at constant rate $\lambda > 0$, the SIR dynamics are

$$\begin{aligned}
\dot{x}_s(t) &= -\beta x_s(t)x_i(t) + \lambda \\
\dot{x}_i(t) &= \beta x_s(t)x_i(t) - \delta x_i(t)
\end{aligned}$$

which we can compactly write in vector form as $\underline{\dot{x}} = \underline{F}(\underline{x})$. The equilibrium point, satisfying $\underline{F}(\underline{x}^*) = \underline{0}$, is $x_s^* = \delta/\beta$ and $x_i^* = \lambda/\delta$. By the Hartman-Grobman theorem, the equilibrium point is locally stable since

$$|z\mathbf{I} - \partial \underline{F}/\partial \underline{x}(\underline{x}^*)| \quad = \quad z^2 + (\beta\lambda/\delta)z + \beta\lambda$$

has roots z with negative real part, where $\mathbf{I}$ is the identity matrix and $|\cdot|$ is here the determinant operator. This approach can be used to establish stability of the typically unique equilibria of the other models considered in this chapter. If the discriminant is positive, i.e., if $\beta\lambda > 4\delta^2$, then the roots are real giving "overdamped" dynamics.

INOCULATION (PATCHING)

Expedited dissemination of patches [146] to remove vulnerabilities to a propagating attack, even so-called "white" worms, can rapidly inoculate susceptible machines. But patches that are not thoroughly tested prior to use can have serious side effects on the operation of the computer and the great variety of software it may be expected to support, e.g., [104]. Moreover, authentication overhead is required to ensure that a patch is legitimate and not itself another form of attack.

Again for simplicity, consider a homogeneous population with $x_i(0) > 0$, $x_s(0) > 0$ and $x_r(0) = 0$. At time t assume individuals decide whether to inoculate/patch with probability $p(x_s(t), x_i(t))$, i.e., we are assuming that "global" information related to x_i and x_s is being at

---

[1]Most of the following models can be easily adjusted to account for exogenous arrivals of infectives too.

least approximately communicated to all susceptible peers (communication which may also involve social dynamics). We assume no further inoculation coordination is taking place.

The (voluntary) inoculation uptake probability $p \in [0, 1]$ is naturally expected to be a decreasing function of the potential severity of side effects of the inoculation/patch, and an increasing function of the potential severity of the disease once infected. For $x_s > 0$, one can also expect p is a continuous, increasing function of the current proportion of the infective population $x_i/(x_i + x_s)$; equivalently, as the proportion of infectives diminishes, there is less incentive to inoculate. Thus, $p(x_s, x_i)$ is increasing in x_i and decreasing in x_s. Moreover, since there is no incentive to inoculate when $x_i = 0$, we assume that for $x_s > 0$:

$$x_i = 0 \quad \Leftrightarrow \quad p(x_s, x_i) = 0. \tag{11.2}$$

Such social dynamics of vaccine uptake can be modeled in greater detail as a game, e.g., [18], and analyzed via Markov decision processes.

The previous assumptions motivate the following real-valued SIR model in continuous time:

$$
\begin{aligned}
\dot{x}_s &= -\beta x_s x_i - p(x_s, x_i)x_s \\
\dot{x}_i &= \beta x_s x_i - \delta x_i \\
\dot{x}_r &= p(x_s, x_i)x_s + \delta x_i,
\end{aligned}
$$

where we have restricted inoculation departures to susceptibles (not infectives). There are two cases for this unforced (closed population) model's equilibrium. If at equilibrium $x_i > 0$, then necessarily $x_s = \delta/\beta$ but then $\dot{x}_s < 0$ so this case cannot hold at equilibrium. Alternatively, if $x_i = 0$ then $p = 0$ so x_s, x_r can take on any values such that $x_s + x_r = N$.

Suppose that susceptibles arrive to the (open) population at a constant rate λ. Also, assume the rate of departure (without return) by movement of individuals in all states x_i, x_s, x_r is governed by the same parameter $\mu > 0$. In this case, the SIR model is:

$$
\begin{aligned}
\dot{x}_s &= -\beta x_s x_i + \lambda - p(x_s, x_i)x_s - \mu x_s \\
\dot{x}_i &= \beta x_s x_i - \delta x_i - \mu x_i \\
\dot{x}_r &= p(x_s, x_i)x_s + \delta x_i - \mu x_r.
\end{aligned}
$$

If we define $X := x_s + x_i + x_r$, then $\dot{X} = \lambda - \mu X$ so that at equilibrium $X = \lambda/\mu$. Also, at equilibrium,

$$
\begin{aligned}
0 &= -(\beta x_i + p(x_s, x_i) + \mu)x_s + \lambda \\
0 &= (\beta x_s - \delta - \mu)x_i.
\end{aligned}
$$

If $x_i = 0$ then $x_s = \lambda/\mu$ (again, since $p(x_s, 0) = 0$ by assumption). Alternatively, if $x_s = (\delta + \mu)/\beta$ in equilibrium then $x_i > 0$ must satisfy:

$$x_i = \frac{\lambda}{\delta + \mu} - \frac{\mu}{\beta} - \frac{1}{\beta}p\left(\frac{\delta + \mu}{\beta}, x_i\right).$$

Assuming $\frac{\lambda}{\delta+\mu} - \frac{\mu}{\beta} > 0$, an equilibrium solution $x_i > 0$ will exist since p is assumed continuous and monotonically increasing in x_i.

11.2 EPIDEMICS SPREAD BY SOCIAL ENGINEERING

Online, sometimes there is a presumption that machines operated by peers that are "socially proximal" are likely to share vulnerability to the same infection ploy whether by:

- hit listing or socially shared software trojans (bond/edge percolation), particularly vulnerable super-peers which are well connected; or

- a compromised website or the social networking site itself (site percolation)[2].

That is, the disease will spread along social ties, unlike some scanning worms discussed in the following section that spread in the network layer. For example, a trojan attachment or URL ploy in a phishing email message more likely will work if there is a social connection between sender and receiver, otherwise it might simply be discarded as spam. For example, the koobface virus spread through social networks [33], and virtual "eggs", used for phishing attacks on Facebook, have exhibited virus-like propagation there. Also, Twitter viruses have recently been reported [122, 140].

Suppose a disease is established at one vertex of such a social graph. Consider a discrete-time disease which spreads along the edges of the graph and infects vertices such that:

1. The disease is established at a given single vertex at time zero.

2. If a vertex v has the disease at time k, then at time $k + 1$:

    (a) some or all (presumed susceptible) uninfected vertices at time k that are connected to v by an edge will be infected; and

    (b) vertex v may be removed from the infective population.

DISCRETE-TIME BRANCHING PROCESS APPROXIMATION

Let X_n be the number of infectives at discrete-time epoch $n \in \mathbb{Z}^+$. For a Galton-Watson branching process, in the next epoch each infective

- will independently generate k infectives with probability p_k for $k \geq 0$, and

- will be removed.

So $\mathsf{E}(X_{n+1}|X_n) = \eta X_n$ where $\eta := \sum_{k \geq 0} k p_k$. The *extinction probability* of the disease $\varepsilon := \mathsf{P}(\lim_{n \to \infty} X_n = 0)$ satisfies

$$\varepsilon = \sum_{k \geq 0} p_k \varepsilon^k.$$

If $\eta < 1$ (respectively, $\eta > 1$) then $\varepsilon = 1$ (respectively, $\varepsilon < 1$). In the "phase transition" example of [56], if $p_0 = (1 - z)^2$, $p_1 = 2z(1 - z)$ and $p_2 = z^2$ for some parameter z satisfying $0 < z < 1$, then $\varepsilon = \min\{(z^{-1} - 1)^2, \ 1\}$.

---

[2]See, e.g., Ch. 3 of [66] for an introduction to percolation.

For infect-and-die SIR (infectives removed after one epoch) acting on a large, though limited, population of N peers: with probability $1 - \varepsilon$, the size of epidemic πN satisfies

$$\pi \;=\; 1 - e^{\pi \eta}.$$

Note that the solution π does not depend on N. Also note that for all $\eta < \infty$, $\pi < 1$ as with the continuous-time SIR model discussed previously.

Now consider a discrete-time disease which spreads along the edges of the graph and infects vertices. The disease is established at a given single vertex at time zero. If a vertex v has the disease at time n, then at time $n + 1$: all or some (presumed susceptible) uninfected vertices at time n that are connected to v by an edge will be infected. It can be shown that a branching process well approximates a discrete-time epidemic on certain random graph models, e.g., Erdos-Renyi (recall Sec. 2.2) and arbitrary degree distribution (see Sec. 3.1 of [53]).

However, the branching process approximation may significantly overestimate the total infected vertex population as the number of uninfected susceptibles diminishes, owing to the increasing probability of re-infection or simultaneous infection attempts [53, 170] (the latter reference on the spread of email virus), recall Sec. 2.2. For many random graph models with sufficiently high edge-degree η (e.g., Erdos-Renyi with $\eta = c + \log N$ for some constant c), it can be shown for large N that complete infection probability has a phase transition when $\eta / \log N = 1$ [53, 56]. Assuming $\eta > \log N$, the *time* till complete infection is at most logarithmic in N by elementary geometric branching. The branching process approximation of [117] can be generalized to consider a random subset/thinning of edges through which disease spreads [115]; see also Sec. 3.5 of [53]. Deterministic (mean field), continuous-time models that correspond to branching process approximations are discussed in [119, 156] and Sec. 4.8 of [53].

EFFECT ON GRAPH TOPOLOGY ON SIS DISEASE SPREAD

Consider a finite graph (V, E) on which a susceptible-infective-susceptible (SIS) disease propagates in continuous time. That is, the vertices can be infected, recover and be infected again repeatedly. The adjacency matrix $\mathbf{A}$ is assumed symmetric, i.e., if $(k, j) \in E$ ($\Leftrightarrow A_{k,j} = 1$) then $(j, k) \in E$. Let X_k indicate whether vertex $k \in V$ is infected. The transition from states 0 to 1 occurs with rate $b \sum_{(k,j)\in E} X_j$ (i.e., infection rate increases with the number of infected neighbors), and from states 1 to 0 at normalized rate 1. Note that the "recovery" could simply be a dormancy strategy intended to evade intrusion detection systems, so that "reinfection" is just reactivation of "observably" infective status. So, in this model, the time τ until the disease terminates (i.e., $X_k = 0$ for all $k \in V$) is finite almost surely.

Claim 11.1   If $b\rho(\mathbf{A}) < 1$ [150] then

$$\mathsf{E}(\tau) \;\le\; \frac{1 + \log n}{1 - b\rho(\mathbf{A})},$$

where $n := |V|$ and $\rho(\mathbf{A})$ is the spectral radius of $\mathbf{A}$.

Sketch of proof[3]: Define another non-negative Markov chain $\underline{Y}$ such that $\underline{Y}(0) = \underline{X}(0)$, and Y_k transitions from integer m to $m+1$ at rate $b \sum_{(k,j)\in E} Y_j$ and from m to $m-1$ at rate $Y_k = m$. Thus, by coupling, $\underline{Y}$ stochastically dominates $\underline{X}$ for all time. Also, by Kolmogorov's forward equation [79], $\frac{d}{dt} E\underline{Y}(t) = (b\mathbf{A} - \mathbf{I})E\underline{Y}(t)$, which implies

$$|E\underline{Y}(t)| = |e^{-(b\mathbf{A}-\mathbf{I})t} E\underline{Y}(0)| \le e^{-(b\rho(\mathbf{A})-1)t}|E\underline{Y}(0)|,$$

where here $|\cdot|$ is Euclidean norm. So,

$$
\begin{aligned}
P(\tau > t) &= P(\sum_k X_k(t) > 0) \\
&\le P(\sum_k Y_k(t) > 0) \\
&\le \sum_k EY_k(t) \\
&\le \sqrt{n}e^{-(b\rho(\mathbf{A})-1)t}|E\underline{Y}(0)|,
\end{aligned}
$$

where the last inequality is Cauchy-Schwarz. Finally, use the identity $E\tau = \int_0^\infty P(\tau > t)dt$. □

For a power law graph whose edge-degree has mean d, maximum m, and distribution with polynomial tail of exponent γ,

$$
\rho(\mathbf{A}) = \begin{cases} (1 + o(1))\sqrt{m} & \text{if } \gamma > 2.5 \\ (1 + o(1))f(\gamma)d^{\gamma-2}m^{3-\gamma} & \text{if } 2.5 > \gamma > 2 \end{cases},
$$

where $f(\gamma) = (\gamma - 2)^{\gamma-1}(\gamma - 1)^{2-\gamma}/(3 - \gamma)$; see Theorem 4 of [34]. Theorem 5.6 of [61] uses the above results to specify sufficient conditions on the parameters γ, d, m of the graph topology and b of the epidemic for slow die-out (large $E\tau$) or fast die-out of the epidemic on the power-law graph.

11.3   ONLINE DISEASE

Online diseases are based on malicious software (herein generically called malcode or malware) that has some type of "reproductive" strategy so that it can move from host computer to host computer. Online diseases have many aspects similar to natural diseases of living populations including:

- certain types of malware can "mutate" to overcome defenses to their spread;

- infected host machines can die (crash); and

---

[3] See Theorem 3.1 of [61].

- a certain vulnerability can be patched (removed) thus, at least partially, inoculating/immunizing the host against future malware that attempts to exploit it (or not patched, resulting in endemic disease).

All computers on the Internet can communicate with each other with great speed. The rate of spread of malware on the Internet can be orders of magnitude faster than the most contagious diseases in humans, even given that geographic hurdles to human disease spread have been overcome by modern methods of travel, particularly by airplane.

In January 2003, the Slammer worm reached and infected 75,000 susceptible SQL servers around the world in less than 20 minutes, and this by random scanning of the IPv4 address space by the collective infectives. Slammer combined the exploit (the "scan" itself) and malware transmission into a single UDP packet. The enormous volume of scanning by Slammer infectives was conspicuous and had significantly disruptive "denial of service" effects on some routers handling the traffic of infected proximal servers. Instead of geographical obstacle to its spread, Slammer's total scan rate was limited by the access bandwidths to the Internet of the domains in which the infectives resided. Slammer automatically spread in the network layer, not using any significant "social" associations.

SIR models were used for Internet worms such as Code Red in [31, 98, 142, 168, 169]. In [84], it was shown how a "stratified" SI model[4] well fit the Slammer worm traces, particularly its rapidly decreasing scan-rate per infective indicating congestion in the domain/enterprise network access links to the core Internet. Recall that for a simple SI model, i.e., with $x_i(t) + x_s(t) = N$ for all time $t \geq 0$, $\dot{x}_i(t) = \beta x_i(N - x_i)$ which has solution [41, 48]:

$$x_i(t) = \frac{x_i(0)N}{x_i(0) + (N - x_i(0))\exp(-\beta Nt)}.$$

Note that $x_i(t) \uparrow N$ as $t \to \infty$.

Now assume the Internet core (connecting peripheral enterprise networks) only negligibly affects any scanning traffic (contagion) which the infectives generate. Consider a population of N enterprise networks. Assume each enterprise has the same number C of susceptible computers. For a homogeneous stratified model, each enterprise is in one of $C + 1$ states where state m connotes exactly m worms (infectives) for $0 \leq m \leq C$. For the entire network, define the state variables $y_m(t)$ representing the number of enterprises in state m at time t.

Clearly, for all time $t \geq 0$, $\sum_{m=0}^{C} y_m(t) = N$ and $x_s := y_0$. Define

$$Y(t) = \sum_{i=1}^{C} y_i(t) = N - y_0(t)$$

as the number of enterprises with one or more worms (infectives). Assume that each such infected enterprise transmits exactly σ scans/s into the Internet irrespective of the "degree" of its infection, i.e., we assume that a single infective saturates the stub-link bandwidth of the enterprise. Finally, an

---

[4]A variation of a "generational" [41, 48] version of the previously described SIR model with $\delta = 0$, i.e., no removals.

implicit assumption is that "local" infections (between nodes in the same enterprise) are negligible in number. Thus, the total rate of scanning (causing infection) into the Internet at time t is

$$S(t) \;=\; \sigma Y(t).$$

The probability that a particular susceptible is infected by a scan is $\eta = 2^{-32}$ (purely random scanning in the 32-bit IPv4 address space). Thus, the probability that a scan causes an enterprise in state m at time t to transition to state $m + 1$ is $(C - m)\eta$ because there are $C - m$ susceptible but not infected nodes in the enterprise at time t. So, define $\beta_m := \sigma\eta(C - m)$. The y_m are governed by the following stratified, though coupled, SI equations: For times $t \geq 0$:

$$\dot{y}_C(t) \;=\; \beta_{C-1}y_{C-1}(t)Y(t), \tag{11.3}$$
$$\dot{y}_m(t) \;=\; (\beta_{m-1}y_{m-1}(t) - \beta_m y_m(t))Y(t) \ \text{ for } 1 \leq m \leq C - 1, \tag{11.4}$$
$$\dot{y}_0(t) \;=\; -\beta_0 y_0(t)Y(t). \tag{11.5}$$

The total number of worms (infectives) at time t is clearly $\sum_{m=1}^{C} my_m(t)$. Thus, the scan-rate per worm is

$$\frac{\sigma Y(t)}{\sum_{m=1}^{C} my_m(t)} \;=\; \frac{\sigma \sum_{m=1}^{C} y_m(t)}{\sum_{m=1}^{C} my_m(t)}.$$

Note that summing equations (11.3) and (11.4) from $m = 1$ to C yields the "standard" SI equation

$$\dot{Y} = \beta_0 y_0 Y = \beta_0 (N - Y)Y$$

whose solution is, again, $Y(t) = Y(0)(1 + \exp(-\beta_0 Nt))^{-1}$ for $t \geq 0$.

A closed-form solution to these equations is derived in [84] together with an explanation of how the (few) parameters were fit to Slammer's empirical data. The system has solution given by

$$y_m(t) \;=\; e^{-(C-m)s(t)} \sum_{j=0}^{m} \binom{C - j}{m - j}(1 - e^{-s(t)})^{m-j} y_j(0)$$

for $m = 0, 1, \ldots, C$, where

$$e^{-s(t)} \;=\; \left(y_0(0) + (1 - y_0(0))e^{\beta Ct}\right)^{-\frac{1}{C}}.$$

The results of Slammer's global activity is depicted in Figure 11.1 [84, 111, 153], together with the (smooth) homogeneous model curves. The oscillations in the empirically measured activity were largely due to measurement errors at the /8 telescope of the University of Wisconsin, which were then magnified by extrapolation, i.e., the scan rates were multiplied by $256 = 2^8$ (see [38] for a discussion of measurement variations of network telescopes).

So, the parsimonious[5], homogeneous stratified SI model overestimated the total scan rate. Better agreement was obtained by a heterogeneous stratified model using concurrent BGP routeview

[5]Just three parameters: N, C, and σ.

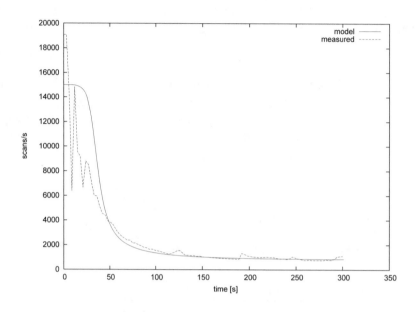

(a) Scan-rate per worm.

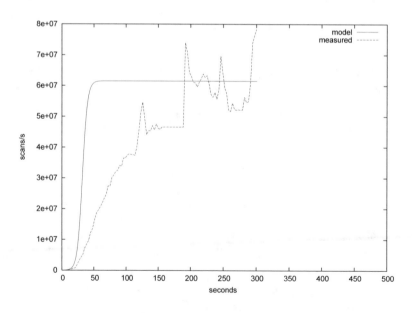

(b) Total scan rate of all worms.

Figure 11.1: The SQL Slammer worm of 2003 and a homogeneous stratified SI model.

information which indicated the *distribution* of susceptible machines per enterprise stub/domain [84]. Note that an extension of the stratified model which could factor the distribution of the domain access bandwidths was also given in [84], although such data was not available to the authors. Figure 11.2 depicts a two-level star topology of the Slammer infectives scaled-down by a factor of 64 to 220 domains, which we used for purposes of simulation on the DETER testbed [153]. The Internet core (presumed ideal switch) is the hub in the middle of the figure.

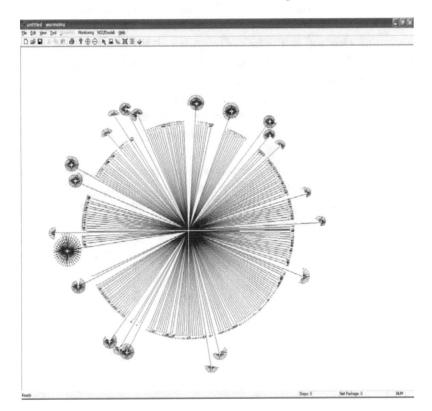

Figure 11.2: A 64:1 scaled-down network using Slammer's routeview data [84].

The bandwidth-limited scanning worm Witty [89] deliberately damaged its infectives, resulting in host crashes during the propagation of the worm. Also, Witty used a less uniform scanning strategy than Slammer. Though not bandwidth limited, the RPC Blaster worm also significantly interfered with the machines it infected by causing them to power down.

11.4 SUMMARY: SELECTING POTENTIAL VICTIMS

Generally, how should the infective y determine the address of another computer x (network layer spreading) which is likely susceptible to y's tactic of infection? One simple scanning strategy is to choose addresses with the same most significant bits as the infective (i.e., to proximal machines in the same domain) under the assumption that if a vulnerability was found in one computer of a given domain, it's likely to exist in others. Also, the local address space of the infective may simply be more populated with machines than on average in the whole Internet. Scanning/vulnerability-probings and infection attempts can be both vertical (port numbers of a single machine) and horizontal (different machine addresses) in nature.

It is well known that a particular automated exploit is often most effective for a "monoculture" of computers that, e.g., use the same version of the same operating system and hence are all likely vulnerable to an exploit targeting it. The lack of operating system monocultures among cell phones was recently used to explain why viruses do *not* spread well through them [148]. This factor can be roughly modeled by the β parameter of the SIR model.

Malware can harvest hit lists of potential victims stored on the host infective, e.g., email addresses or account identifiers of a social networking site visited by the host (application layer spreading). Such social associations may indicate a greater propensity to be infectable by the same ploys.

Finally, we mention that malware, particularly botnets like Conficker, may spread using a variety of sophisticated, covert techniques [113, 128]. A primary mode of spread of the now propagating and notorious Stuxnet worm is via an *autorun.inf* file in the root directory of USB memory sticks [167], i.e., spreading via physical human-to-computer contact.

Bibliography

[1] Z. Abrams, R. McGrew, and S. Plotkin. A non-manipulable trust system based on eigentrust. *ACM SIGecom Exchanges*, 5(4), July 2005. DOI: 10.1145/1120717.1120721 60

[2] D. Acemoglu, M.A. Dahleh, I. Lobel, and A. Ozdaglar. Bayesian learning in social networks. *preprint available at* http://econ-www.mit.edu/files/5201, Jan. 2010. 75

[3] A. Al-Nowaihi and P.L. Levine. The stability of the Cournot oligopoly model: A reassessment. *Journal of Economic Theory*, 35:307–321, 1985. DOI: 10.1016/0022-0531(85)90046-8 29

[4] D. Aldous and J. Fill. Reversible Markov Chains and Random Walks on Graphs. http://www.stat.berkeley.edu/~aldous/RWG/book.html. 51

[5] T. Alpcan, T. Basar, and S. Dey. A power control game based on outage probabilities for multicell wireless data networks. In *Proc. IEEE ACC*, 2004. DOI: 10.1109/TWC.2006.1618938 32

[6] T. Alpcan, T. Basar, R. Srikant, and E. Altman. CDMA uplink power control as a noncooperative game. *Wireless Networks*, 8, Nov. 2002. DOI: 10.1023/A:1020375225649 32

[7] E. Altman, T. Boulogne, R. El-Azouzi, T. Jiménez, and L. Wynter. A survey on networking games in telecommunications. *Comput. Oper. Res.*, 33(2):286–311, 2006. DOI: 10.1016/j.cor.2004.06.005 27

[8] R.M. Anderson and R.M. May. *Infectious Disease of Humans: Dynamics and Control*. Oxford University Press, Oxford, 1991. 87

[9] P. Antoniadis, C. Courcoubetis, and R. Mason. Comparing economic incentives in peer-to-peer networks. *Computer Networks*, 46(1):133–146, 2004. DOI: 10.1016/j.comnet.2004.03.021 65

[10] A. Aram, C. Singh, S. Sarkar, and A. Kumar. Cooperative profit sharing in coalition based resource allocation in wireless networks. In *Proc. IEEE INFOCOM*, Rio de Janeiro, Brazil, April 2009. DOI: 10.1109/INFCOM.2009.5062136 30

[11] J. Aspnes. Randomized protocols for asynchronous consensus. *Distributed Computing*, 16(2-3):165–175, Sept. 2003. DOI: 10.1007/s00446-002-0081-5 83

[12] J. Aspnes, Z. Diamadi, and G. Shah. Fault-tolerant routing in peer-to-peer systems. In *Proc. ACM Symp. Principles of Distributed Computing*, 2002. DOI: 10.1145/571825.571862 42

[13] H. Attiya and J. Welch. *Distributed Computing: Fundamentals, Simulations, and Advanced Topics, 2nd Ed.* Wiley, New York, 2004. 82, 83

[14] A. P. Azad, E. Altman, and R. Elazouzi. From altruism to non-cooperation in routing games. In *Networking and Electronic Commerce Research Conference (NAEC)*, Lake Garda, Italy, Oct. 2009. 30

[15] F. Baccelli and B. Blaszczyszyn. *Stochastic Geometry and Wireless Networks, Part I: Theory.* NOW, 2009. 25

[16] A. Bar-Noy, D. Dolev, C. Dwork, and R. Strong. Shifting gears: Changing algorithms on the fly to expedite Byzantine agreement. In *Proc. 6th Annual ACM Symp. Principles Distributed Computing*, pages 42 – 51, 1987. DOI: 10.1145/41840.41844 79

[17] T. Basar and G. J. Olsder. *Dynamic noncooperative game theory, 2nd Ed.* Academic Press, 1995. 29

[18] C.T. Bauch and D.J.D. Earn. Vaccination and the theory of games. *Proc. Nat'l Acad. Sci.*, 101:13391–13394, 2004. DOI: 10.1073/pnas.0403823101 88

[19] D.P. Bertsekas and J. N. Tsitsiklis. Convergence rate and termination of asynchronous iterative algorithms. In *Proc. 3rd International Conference on Supercomputing*, 1989. DOI: 10.1145/318789.318894 29

[20] D. Bickson and D. Malkhi. A unifying framework for rating users and data items in peer-to-peer social networks. *Peer-to-Peer Networking and its Applications Journal*, April 2008. DOI: 10.1007/s12083-008-0008-4 77

[21] J. Bieber, M. Kenney, N. Torre, and L.P. Cox. An empirical study of seeders in BitTorrent. Technical Report CS-2006-08, Duke University, Computer Science Dept., 2006. 67, 70

[22] BitTorrent. http://www.bittorrent.com. 65

[23] B. Bollobas. *Random Graphs, 2nd Ed.* Cambridge University Press, Cambridge, 2001. 18

[24] B. Bollobas and O. Riordan. Robustness and vulnerability of scale-free random graphs. *Internet Mathematics*, 1(1), 2003. 25

[25] S. Bornholdt and H. Ebel. World Wide Web scaling exponent from Simon's 1955 model. *Physical Review E*, 64:035104–1, 2001. DOI: 10.1103/PhysRevE.64.035104 20

[26] S. Boyd, A. Ghosh, B. Prabhakar, and D. Shah. Randomized gossip algorithms. *IEEE Trans. Info. Theory*, 52(6), June 2006. DOI: 10.1109/TIT.2006.874516 60, 75, 76

[27] S. Brooks, A. Gelman, G. Jones, and X.-L. Meng. *Handbook of Markov Chain Monte Carlo.* Chapman & Hall/CRC Press, 2010. 54

[28] G.W. Brown. Iterative solutions of games with fictitious play. In T.C. Koopmans, editor, *Activity Analysis of Production and Allocation*, Wiley, New York, 1951. 28

[29] S. Caron, G. Kesidis, and E. Altman. Application neutrality and a paradox of side payments. In *Proc. ACM Re-Architecting the Internet (ReArch) Workshop*, Philadelphia, Nov. 2010. 35

[30] M. Castro, P. Druschel, A. Ganesh, A. Rowstron, and D.S. Wallach. Security for structured peer-to-peer overlay networks. In *Proc. USENIX OSDI*, Boston, Dec. 2002. DOI: 10.1145/1060289.1060317 61

[31] Z. Chen, L. Gao, and K. Kwait. Modeling the spread of active worms. In *Proc. IEEE INFOCOM*, San Francisco, 2003. DOI: 10.1109/INFCOM.2003.1209211 92

[32] A. Cheng and E. Friedman. Sybilproof reputation mechanisms. In *Proc. ACM SIGCOMM P2PECON workshop*, pages 128–132, 2005. DOI: 10.1145/1080192.1080202 58, 62

[33] E. Chien and J. Shearer. W32.Koobface. `http://www.symantec.com/security_response/writeup.jsp?docid=2008-080315-0217-99`, Apr.-10-2010. 89

[34] F. Chung, L. Lu, and V. Vu. Eigenvalues of random power law graphs. *Annals of Combinatorics*, 7, 2003. 91

[35] I. Clarke, S.G. Miller, O. Sandberg, B. Wiley, and T.W. Hong. Protecting freedom of information online with Freenet. *IEEE Internet Computing*, Jan. 2002. DOI: 10.1109/4236.978368 49

[36] B. Cohen. Incentives Build Robustness in BitTorrent. In *Workshop on Economics of Peer-to-Peer Systems*, Berkeley, CA, May 2003. 65

[37] E. Cohen and S. Shenker. Replication strategies in unstructured peer-to-peer networks. In *Proc. ACM SIGCOMM*, Aug. 2002. DOI: 10.1145/964725.633043 13

[38] E. Cooke, M. Bailey, Z.M. Mao, D. Watson, F. Jahanian, and D. McPherson. Toward understanding distributed blackhole placement. In *Proc. ACM WORM*, Washington, D.C., Oct. 29, 2004. DOI: 10.1145/1029618.1029627 93

[39] Comcast Corporation. Description of planned network management practices to be deployed following the termination of current practices, Attachment B. available at `http://downloads.comcast.net/docs/Attachment_B_Future_Practices.pdf`. 8

[40] Y.K. Dalal and R.M. Metcalfe. Reverse path forwarding of broadcast packets. *Communications of the ACM*, vol. 21, no. 12:p. 1040–1048, Dec. 1978. DOI: 10.1145/359657.359665 47

[41] D.J. Daley and J. Gani. *Epidemic Modeling: An Introduction*. Cambridge University Press, Cambridge, 1999. DOI: 10.1017/CBO9780511608834 68, 85, 86, 92

[42] G. Danezis and P. Mittal. SybilInfer: Detecting Sybil Nodes using Social Networks. In *Proc. NDSS*, 2009. 61

[43] D. DeFigueiredo and E.T. Barr. TrustDavis: A non-exploitable online reputation system. In *Proc. IEEE CEC*, July 2005. DOI: 10.1109/ICECT.2005.98 62

[44] M. Dell'Amico and L. Capra. SOFIA: Social filtering for robust recommendations. In *Proc. IFIPTM*, 2008. DOI: 10.1007/978-0-387-09428-1_9 60

[45] M. Dell'Amico and Y. Roudier. A measurement of mixing time in social networks. In *Proc. Int'l Workshop on Security and Trust Management (STM)*, Saint Malo, France. Elsevier, Sept. 2009. 22, 25, 52, 53, 61

[46] A. Dembo and O. Zeitouni. *Large Deviations Techniques and Applications*. Jones and Bartlett, Boston, 1992. 54

[47] P. Diaconis and D. Stroock. Geometric bounds for eigenvalues of Markov chains. *Annals of Applied Probability*, 1(1), Feb. 1991. DOI: 10.1214/aoap/1177005980 51, 52

[48] O. Diekmann and J.A.P. Heesterbeek. *Mathematical epidemiology of infectious diseases: Model building, analysis, and interpretation*. Wiley, 2000. 68, 69, 85, 86, 92

[49] S.N. Dorogovtsev and J.F.F. Mendes. Evolution of networks. *Adv. Phys*, 51:1079–1187, 2002. DOI: 10.1080/00018730110112519 18, 20, 22

[50] J.R. Douceur. The sybil attack. In *Proc. First Int'l Workshop on Peer-to-Peer Systems*, London, 2004. 61

[51] R.O. Duda, P.E. Hart, and D.G. Stork. *Pattern Classification (2nd ed)*. Wiley, 2000. 17

[52] R. Durrett. *The Essentials of Probability*. Duxbury Press, Belmont, CA, 1994. xiii

[53] R. Durrett. *Random Graph Dynamics*. Cambridge University Press, New York, 2007. 18, 19, 20, 23, 77, 90

[54] N. Economides. Net neutrality: Non-discrimination and digital distribution of content through the Internet. *I/S: A Journal of Law and Policy*, 4(2):209–233, 2008. 35

[55] P. Erdos and A. Renyi. The evolution of random graphs. *Magyar Tud. Akad. Mat. Kutato Int. Kozl.*, 5:17–61, 1960. 22

[56] P.T. Eugster, R. Guerraoui, A.-M. Kermarrec, and L. Massoulie. Epidemic information dissemination in distributed systems. *IEEE Computer*, 37:60–67, 2004. DOI: 10.1109/MC.2004.1297243 85, 89, 90

[57] M. Feldman, K. Lai, I. Stoica, and J. Chuang. Robust incentive techniques for peer-to-peer networks. In *5th ACM conference on Electronic commerce*, pages 102 – 111, New York, 2004. DOI: 10.1145/988772.988788 25

[58] W. Feller. *An Introduction to Probability Theory and its Applications*. Wiley, New York, 1968. 9

[59] L.C. Freeman. A set of measures of centrality based on betweenness. *Sociometry*, 40(1):35–41, March 1977. DOI: 10.2307/3033543 24

[60] J.W. Friedman and C. Mezzetti. Learning in games by random sampling. *Journal of Economic Theory*, 98(1):55–84, May 2001. DOI: 10.1006/jeth.2000.2694 28

[61] A. Ganesh, L. Massoulie, and D. Towsley. The effect of network topology on the spread of epidemics. In *in Proc. IEEE INFOCOM*, Mar. 2005. DOI: 10.1109/INFCOM.2005.1498374 91

[62] J.A. Garay and Y. Moses. Fully polynomial Byzantine agreement for $n > 3t$ processors in $t + 1$ rounds. *SIAM J. Comput.*, 27, 1998. DOI: 10.1137/S0097539794265232 82

[63] C. Gkantsidis, M. Mihail, and A. Saberi. Random walks in peer-to-peer networks. In *Proceedings of IEEE INFOCOM*, 2004. DOI: 10.1016/j.peva.2005.01.002 51

[64] B. Golub and M.O. Jackson. How homophily affects learning and diffusion in networks. Technical Report arXiv:0811.4013v1, arxiv.org, Jan. 29, 2009. 24, 51, 52

[65] C. Griffin, G. Kesidis, P. Antoniadis, and S. Fdida. File-sharing segment distribution: acquisition performance, content availability, and cooperation incentive. *preprint*, 2010. 70

[66] G.R. Grimmett. Probability on graphs. `http://www.statslab.cam.ac.uk/~grg/books/pgs.pdf`, 2010. 23, 77, 78, 89

[67] G.R. Grimmett and D.R. Stirzaker. *Probability and Random Processes, Third Edition*. Oxford University Press, 2001. xiii, 51, 77

[68] V. Guruswami. Rapidly mixing Markov chains: A comparison of techniques. `http://www.cs.washington.edu/homes/venkat/papers/markov-survey.ps`, 2000. 51, 53

[69] R. Holley and D. Stroock. Simulated annealing via Sobolev inequalities. *Communications in Mathematical Physics*, Vol. 115, No. 4:pp. 553–569, Sept. 1988. DOI: 10.1007/BF01224127 78

[70] J.J. Hopfield. Neural networks and physical systems with emergent collective computational abilities. *Proc. Nat'l Academy of Sciences of the USA*, 79(8):2554–2558, April 1982. DOI: 10.1073/pnas.79.8.2554 76

[71] R.A. Horn and C.R. Johnson. *Matrix Analysis*. Cambridge University Press, 1985. 10, 52, 60

[72] A.S. Householder. *Matrices in Numerical Analysis*. Dover, 1986. 60

[73] Y. Jin and G. Kesidis. Equilibria of a noncooperative game for heterogeneous users of an ALOHA network. *IEEE Communications Letters*, Vol. 6, No. 7:pp. 282–284, 2002. DOI: 10.1109/LCOMM.2002.801326 28, 29

[74] Y. Jin and G. Kesidis. A pricing strategy for an ALOHA network of heterogeneous users with inelastic bandwidth requirements. In *CISS, Princeton*, March 2002. 28, 29

[75] Y. Jin and G. Kesidis. Dynamics of usage-priced communication networks: the case of a single bottleneck resource. *IEEE/ACM Trans. Networking*, Oct. 2005. DOI: 10.1109/TNET.2005.857120 30

[76] S.D. Kamvar, M.T. Schlosser, and H. Garcia-Molina. The eigentrust algorithm for reputation management in P2P networks. In *Proc. WWW*, pages 640–651, 2003. DOI: 10.1145/775152.775242 57, 58, 59, 60

[77] D. Karger, E. Lehman, T. Leighton, R. Panigrahy, M. Levine, and D. Lewin. Consistent hashing and random trees: distributed caching protocols for relieving hot spots on the world wide web. In *Proc. ACM STOC*, 1997. DOI: 10.1145/258533.258660 13, 14, 15

[78] D. Karger and M. Ruhl. Simple efficient load balancing algorithms for peer-to-peer systems. In *Proc. 16th ACM Symposium on Parallelism in Algorithms and Architectures (SPAA)*, 2004. DOI: 10.1145/1007912.1007919 15, 43, 47

[79] S. Karlin and H.M. Taylor. *A First Course in Stochastic Processes, 2nd Ed.* Academic Press Inc., New York, 1975. xiii, 9, 77, 91

[80] R. Karp. Probabilistic recurrence relations. *Journal of the ACM*, 41(6):1136–1150, Nov. 1994. DOI: 10.1145/195613.195632 42

[81] R.M. Karp, E. Upfal, and A. Wigderson. The complexity of parallel search. *Journal of Computer and System Sciences*, 36(2):225–253, 1988. DOI: 10.1016/0022-0000(88)90027-X 42

[82] F.P. Kelly. The mathematics of traffic in networks. In T. Gowers; J. Barrow-Green and I. Leader (assoc.), editors, *The Princeton Companion to Mathematics*, pages 862–870. Princeton University Press, Princeton, NY, 2008. 8

[83] G. Kesidis, Y. Jin, E. Altman, and A. Azad. Stable Nash equilibria of medium access games under symmetric, socially altruistic behavior. Technical Report arXiv:1003.5324, arxiv.org, March 27, 2010. 29

[84] G. Kesidis, M. Vojnovic, I. Hamadeh, Y. Jin, and S. Jiwasurat. Model of the spread of randomly scanning internet worms that saturate access links. *ACM TOMACS*, April 2008. DOI: 10.1145/1346325.1346327 92, 93, 95

[85] R. Kinderman and J.L. Snell. *Markov Random Fields and their Applications*. AMS, 1980. 77, 78

[86] J. Kleinberg. The small-world phenomenon: an algorithmic perspective. In *Proc. 32nd ACM Symp. on Theory of Computing*, Portland, OR, 2000. DOI: 10.1145/335305.335325 42, 45

[87] T. Konstantopoulos and G. Kesidis. Randomized data object indexing for structured search. Technical Report CSE-08-007, Penn State Univ, CSE Dept, Apr. 2008, available at http://www.cse.psu.edu/~kesidis/book/beta.pdf. 43

[88] T. Konstantopoulos, G. Kesidis, and P. Sousi. A stochastic epidemiological model and a deterministic limit for BitTorrent-like peer-to-peer file-sharing networks. In *Proc. Workshop on Network Control and Optimization (NET-COOP), Paris*, volume Springer LNCS 5425, Sept. 2008. DOI: 10.1007/978-3-642-00393-6_4 69, 86

[89] A. Kumar, V. Paxson, and N. Weaver. Exploiting underlying structure for detailed reconstruction of an internet-scale event. In *Proceedings of ACM IMC*, 2005. 95

[90] A. Kumar, J. Xu, and E.W. Zegura. Efficient and scalable query routing for unstructured peer-to-peer networks. In *Proc. IEEE INFOCOM*, 2005. DOI: 10.1109/INFCOM.2005.1498343 14, 60

[91] J.F. Kurose and K.W. Ross. *Computer Networking: A Top-Down Approach Featuring the Internet*. Addison-Wesley, Boston, 2002. xiii, 3, 16

[92] T. Kurtz. *Approximation of Population Processes*. SIAM, Philadelphia, 1981. 86

[93] L. Lamport, R. Shostak, and M. Pease. The Byzantine Generals Problem. *ACM Transactions on Programming Languages and Systems*, 4:382–401, 1982. DOI: 10.1145/357172.357176 79

[94] J. Ledlie, L. Serban, and D. Toncheva. Scaling filename queries in a large-scale distributed file system. Technical Report TR-03-02, Harvard University, Jan. 2002. 14, 60

[95] A. Legout, A. Liogkas, E. Kohler, and L. Zhang. Clustering and sharing incentives in BitTorrent systems. In *Proc. ACM SIGMETRICS*, San Diego, CA, June 2007. DOI: 10.1145/1269899.1254919 65, 69

[96] C. Lesniewski-Lass. A Sybil-proof one-hop DHT. In *Proc. ACM SocialNets*, Glasgow, Scotland, Apr. 2008. DOI: 10.1145/1435497.1435501 60

[97] D.A. Levin, Y. Peres, and E.L. Wilmer. *Markov Chains and Mixing Times*. AMS, Providence, RI, 2008. 25, 51

[98] M. Liljenstam, D. Nicol, V. Berk, and R. Gray. Simulating realistic network worm traffic for worm warning system design and testing. In *Proc. ACM WORM*, Washington, D.C., 2003. DOI: 10.1145/948187.948193 92

[99] J. Liu and J. Xu. Proxy caching for media streaming over the Internet. *IEEE Communications Mag.*, 42(8), Aug. 2004. DOI: 10.1109/MCOM.2004.1321397 11

[100] Q. Lv, P. Cao, E. Cohen, K. Li, and S. Shenker. Search and replication in unstructured peer-to-peer networks. In *Proc. International Conference on Supercomputing (ICS)*, 2002. DOI: 10.1145/514191.514206 13

[101] P. Massa and P. Avesani. Trust aware recommender systems. In *Proc. ACM Recommender Systems Conference*, Minneapolis, 2007. DOI: 10.1145/1297231.1297235 60

[102] L. Massoulie and M. Vojnovic. Coupon replication systems. *IEEE/ACM Transactions on Networking*, Vol. 16, No. 3, June 2008. DOI: 10.1145/1071690.1064215 68

[103] P. Maymounkov and D. Mazieres. Kademlia: A peer-to-peer information system based on the XOR metric. In *Proceedings of IPTPS*, Cambridge, March 2002. DOI: 10.1007/3-540-45748-8_5 42

[104] D. McCullagh. Buggy McAfee update whacks Windows XP PCs. `http://news.cnet.com/8301-1009_3-20003074-83.html`, 2010. 87

[105] J. McLachlan, A. Tan, N. Hopper, and Y. Kim. Scalable onion routing with Torsk. In *Proc. ACM CCS*, Chicago, Nov. 2009. DOI: 10.1145/1653662.1653733 49

[106] I. Menache and N. Shimkin. Rate-based equilibria in collision channels with fading. *IEEE JSAC*, 26(7):1070–1077, Sept. 2008. DOI: 10.1109/JSAC.2008.080905 32, 34

[107] F. Meshkati, M. Chiang, H.V. Poor, and S.C. Schwartz. A game-theoretic approach to energy-efficient power control in multicarrier CDMA systems. *IEEE JSAC*, 24(6), June 2006. DOI: 10.1109/JSAC.2005.864028 32

[108] M. Mitzenmacher and E. Upfal. *Probability and Computing: Randomized Algorithms and Probabilistic Analysis*. Cambridge University Press, New York, 2005. 51

[109] Michael Mitzenmacher. A brief history of generative models for power law and lognormal distributions. *Internet Mathematics*, 1(2):226–251, 2004. 18

[110] D. Mollison. The structure of epidemic models. In *Epidemic Models: Their Structure and Relation to Data*, Cambridge University Press, Cambridge, UK, 1995. DOI: 10.2277/0521475368 87

[111] D. Moore, V. Paxson, S. Savage, C. Shannon, S. Staniford, and N. Weaver. Inside the Slammer worm. *IEEE Security and Privacy*, 2004. DOI: 10.1109/MSECP.2003.1219056 93

[112] B. Mortazavi and G. Kesidis. Incentive-compatible cumulative reputation systems for peer-to-peer file-swapping. In *Proc. CISS*, Princeton, NJ, March 2006. 58

[113] J. Nazario and T. Holz. As the net churns: Fast-flux botnet observations. In *Proc. 3rd Int'l Conf. on Malicious and Unwanted Software (MALWARE)*, 2008.
DOI: 10.1109/MALWARE.2008.4690854 96

[114] A. Neishaboori and G. Kesidis. Routing and uplink-downlink scheduling in ad hoc CDMA networks. In *Proc. IEEE ICC*, Istanbul, June 2006. DOI: 10.1109/ICC.2006.254826 32

[115] M.E.J. Newman. Spread of epidemic disease on networks. *Phys. Rev. E*, 66, 2002.
DOI: 10.1103/PhysRevE.66.016128 19, 90

[116] M.E.J. Newman. Detecting communication structure in networks. *Eur. Phys. J. B*, 38(2):321–330, 2004. DOI: 10.1140/epjb/e2004-00124-y 25, 54

[117] M.E.J. Newman, S.H. Strogatz, and D.J. Watts. Random graphs with arbitrary degree distributions and their applications. *Phys. Rev. E*, 64, 2001. DOI: 10.1103/PhysRevE.64.026118
19, 90

[118] G. Palla, I. Derenyi, I. Farkas, and T. Vicsek. Uncovering overlapping community structure of complex networks in nature and society. *Nature*, 435(7043):814–818, 2005.
DOI: 10.1038/nature03607 25, 54

[119] R. Pastor-Satorras and A. Vespignani. Epidemic spreading in scale-free networks. *Phys. Rev. Let.*, 86(14):3200–3203, 2001. DOI: 10.1103/PhysRevLett.86.3200 90

[120] P. Patankar, G. Nam, G. Kesidis, and C. Das. Exploring anti-spam models in VoIP systems. In *Proc. IEEE ICDCS*, Beijing, 2008. DOI: 10.1109/ICDCS.2008.71 60, 64

[121] P. Patankar, G. Nam, G. Kesidis, T. Konstantopoulos, and C. Das. Peer-to-peer unstructured anycasting using correlated swarms. In *Proc. ITC*, Paris, Sept 2009. 43, 54

[122] I. Paul. Twitter Worm: A Closer Look at What Happened. http://www.pcworld.com/article/163054/twitter_worm_a_closer_look_at_what_happened.html, Apr. 14, 2009. 89

[123] R. Pemantle. Nonconvergence to unstable points in urn models and stochastic approximations. *Annals of Probability*, 18:698 – 712, 1990. DOI: 10.1214/aop/1176990853 29

[124] L.L. Peterson and B.S. Davie. *Computer Networks, A System Approach, 2nd Ed.* Morgan Kaufmann, 2000. xiii, 3

[125] C. Piro, C. Shields, and B. N. Levine. Detecting the sybil attack in mobile ad hoc networks. In *Proc. IEEE/ACM SecureComm*, Aug. 2006. DOI: 10.1109/SECCOMW.2006.359558 61

[126] J.G. Proakis. *Communication Systems Engineering*. Prentice Hall International Editions, 1994. 33

[127] D. Qiu and R. Srikant. Modeling and performance analysis of BitTorrent-like peer-to-peer networks. In *Proc. ACM SIGCOMM*, Portland, OR, 2004. DOI: 10.1145/1030194.1015508 65, 69

[128] J. Riden. Know your enemy: Fast-flux service networks. Available at: http://www.honeynet.org/node/132, Aug. 16, 2008. 96

[129] K.W. Ross. Hash-routing for collections of shared web caches. *IEEE Network*, 11:37–44, 1997. DOI: 10.1109/65.642358 10, 11

[130] K.W. Ross and D. Rubenstein. Tutorial on P2P systems. Available at http://cis.poly.edu/~ross/papers/P2PtutorialInfocom.pdf, 2004. 13

[131] T. Roughgarden. *Selfish routing and the price of anarchy*. MIT Press, Cambridge, MA, 2005. 35

[132] S. Roy and J. Zhu. A 802.11 based slotted dual-channel reservation MAC protocol for in-building multi-hop networks. *Mobile Networks and Applications*, 10(5), Oct. 2005. DOI: 10.1145/1160143.1160145 32

[133] S. Ruohomaa, L. Kutvonen, and E. Koutrouli. Reputation management survey. In *Proc. Int'l Conf. on Reliability and Security (ARES)*, 2007. DOI: 10.1109/ARES.2007.123 60

[134] S. Sanghavi, B. Hajek, and L. Massoulie. Gossiping with multiple messages. *IEEE Trans. Info. Th.*, 53(12):4640–4654, 2007. DOI: 10.1109/TIT.2007.909171 75

[135] C.U. Saraydar, N.B. Mandayam, and D.J. Goodman. Pareto efficiency of pricing based control in wireless data networks. In *IEEE WCNC*, pages pp. 231–234, 1999. DOI: 10.1109/WCNC.1999.797821 32

[136] C.U. Saraydar, N.B. Mandayam, and D.J. Goodman. Power control in a multi-cell CDMA data system using pricing. In *IEEE VTC*, pages pp. 484–491, 2000. DOI: 10.1109/VETECF.2000.887063 32

[137] Top applications (bytes) for subinterface 0[0]: SD-NAP traffic. http://www.caida.org/analysis/workload/byapplication/sdnap/. 65

[138] J. Seade. The stability of Cournot revisited. *Journal of Economic Theory*, 23:15–27, 1980. DOI: 10.1016/0022-0531(80)90028-9 29

[139] H. Simon. On a class of skew distribution functions. *Biometrika*, 42(3/4):425–440, 1955. DOI: 10.2307/2333389 20, 21

[140] spamNEWS administrator. Twitter Destroys 2nd PC Worm During Same Week. `http://www.spamfighter.com/News-15157-Twitter-Destroys-2nd-Pc-Worm-During-Same-Week.htm`, Oct. 4, 2010. 89

[141] M. Spear, X. Lu, N. Matloff, and S.F. Wu. A formal model to analyze and compare reputation systems for distributed networks. In *Proc. IEEE INFOCOM,* San Diego, 2010. 60

[142] S. Staniford, V. Paxson, and N. Weaver. How to own the Internet in your spare time. In *Proc. USENIX Security Symposium*, pages 149–167, Aug. 2002. 92

[143] A.-J. Su, D. Choffnes, A. Kuzmanovic, and F. Bustamante. Drafting behind Akamai (Travelocity-based detouring). In *Proc. ACM SIGCOMM*, Pisa, Sept. 2006. DOI: 10.1145/1151659.1159962 11

[144] W.W. Tepstra, J. Kangasharju, C. Leng, and A.P. Buchmann. BubbleStorm: Resilient, probabilistic, and exhaustive peer-to-peer search. In *Proc. ACM SIGCOMM*, Kyoto, Aug. 2007. DOI: 10.1145/1282427.1282387 51

[145] utorrent. `http://www.utorrent.com`. 7

[146] M. Vojnovic and A. Ganesh. On the effectiveness of automatic patching. In *Proc. ACM Workshop on Rapid Malcode (WORM)*, pages 41–50, 2005. DOI: 10.1145/1103626.1103634 87

[147] B. Walsh. Markov Chain Monte Carlo and Gibbs Sampling. `http://membres-timc.imag.fr/Olivier.Francois/mcmc_gibbs_sampling.pdf`, 2004. 78

[148] P. Wang, M.C. Gonzales, C.A. Hidalgo, and A.-L. Barabasi. Understanding the spreading patterns of mobile phone viruses. *Science*, Apr. 2009 (online). DOI: 10.1126/science.1167053 96

[149] P. Wang, I. Osipkov, N. Hopper, and Y. Kim. Secure and robust DHT routing. Technical Report 2006/20, University of Minnesota DTC, 2006. 49

[150] Y. Wang, D. Chakrabarti, C. Wang, and C. Faloutsos. Epidemic spreading in real networks: An eigenvalue viewpoint. In *Proc. IEEE SRDS*, 2003. DOI: 10.1109/RELDIS.2003.1238052 90

[151] D.J. Watts, P.S. Dodds, and M.E.J. Newman. Identity and search in social networks. *Science*, 296:1302–1305, May 17, 2002. DOI: 10.1126/science.1070120 25, 54

[152] D.J. Watts and S.H. Strogatz. Collective dynamics of 'small-world' networks. *Nature*, Vol. 393, 1998. DOI: 10.1038/25379 22

[153] N. Weaver, I. Hamadeh, G. Kesidis, and V. Paxson. Preliminary results using scale-down using scale-down to explore worm dynamics. In *Proc. ACM WORM*, Washington, D.C., Oct. 2004. DOI: 10.1145/1029618.1029628 93, 95

[154] J.W. Weibull. *Evolutionary Game Theory*. MIT Press, Cambridge, MA, 1995. 28, 29

[155] Wikipedia. The beta distribution.
http://en.wikipedia.org/wiki/Beta_distribution. 47

[156] J.R. Wilson. Note on "influences of resource limitations and transmission costs on epidemic simulations and critical thresholds in scale-free networks". *to appear in Simulation: Transactions of the Society for Modeling and Simulation International*, 2009. DOI: 10.1177/0037549710366018 90

[157] R.W. Wolff. *Stochastic Modeling and the Theory of Queues*. Prentice-Hall, Englewood Cliffs, NJ, 1989. xiii, 12, 46, 67

[158] M. Xiao, N.B. Shroff, and E.K.P. Chong. A utility-based power-control scheme in wireless cellular systems. *IEEE/ACM Trans. Networking*, 11, April 2003. DOI: 10.1109/TNET.2003.810314 32

[159] H. Xie, Y.R. Yang, A. Krishnamurthy, Y. Liu, and A. Silberschatz. P4P: Portal for Applications. In *Proc. ACM SIGCOMM*, Seattle, Aug. 2008. DOI: 10.1145/1402946.1402999 11

[160] M. Yannuzzi, X. Masip-Bruin, S. Sanchez, J. Domingo-Pascual, A. Orda, and A. Sprintson. On the challenges of establishing disjoint QoS IP/MPLS paths across multiple domains. *IEEE Communications Magazine*, Dec. 2006. DOI: 10.1109/MCOM.2006.273101 6

[161] S. Yi, X. Deng, G. Kesidis, and C.R. Das. A dynamic quarantine scheme for controlling unresponsive TCP flows. *Telecommunication Systems*, Vol. 37, No. 4, 2008. DOI: 10.1007/s11235-008-9104-2 8

[162] H. Yu, P.B. Gibbons, M. Kaminsky, and F. Xiao. SybilLimit: A Near-Optimal Social Network Defense against Sybil Attacks. In *IEEE Symposium on Security and Privacy*, 2008. DOI: 10.1109/TNET.2009.2034047 60, 62

[163] H. Yu, M. Kaminsky, P.B. Gibbons, and A. Flaxman. SybilGuard: Defending against sybil attacks via social networks. In *ACM SIGCOMM*, pages 267–278, 2006. DOI: 10.1145/1151659.1159945 61, 62

[164] A. Zhang and Y. Zhang. Stability of Nash equilibrium: The multiproduct case. *Journal of Mathematical Economics*, 26(4):441–462, 1996. DOI: 10.1016/0304-4068(95)00760-1 29

[165] H. Zhang, A. Goel, and R. Govindan. Using the small-world model to improve Freenet performance. *Computer Networks Journal*, vol. 46, no. 4:555–574, Nov. 2004. DOI: 10.1016/j.comnet.2004.05.004 49

[166] M. Zhong and K. Shen. Popularity-biased random walks for peer-to-peer search under the square-root principle. In *Proc. 5th International Workshop on Peer-to-Peer Systems (IPTPS)*, Santa Barbara, CA, Feb. 2006. 13, 53

[167] Z. Zorz. Examining the Stuxnet worm. `http://www.net-security.org/malware_news.php?id=1471`. 96

[168] C.C. Zou, W. Gong, and D. Towsley. Code Red worm propagation modeling and analysis. In *Proc. 9th ACM Conference on Computer and Communication Security (CCS'02)*, Washington, D.C., Nov. 2002. DOI: 10.1145/586110.586130 92

[169] C.C. Zou, W. Gong, and D. Towsley. Worm propagation modeling and analysis under dynamic quarantine defense. In *ACM CCS Workshop on Rapid Malcode (WORM'03)*, Washington, D.C., Oct. 2003. DOI: 10.1145/948187.948197 92

[170] C.C. Zou, D. Towsley, and W. Gong. Modeling and simulation study of the propagation and defense of Internet email worm. *IEEE Trans. on Dependable and Secure Computing*, 4(2):105–118, April-June 2007. DOI: 10.1109/TDSC.2007.1001 19, 25, 90

Author's Biography

GEORGE KESIDIS

George Kesidis received his B.A.Sc. in Electrical Engineering from the University of Waterloo in 1988, and his M.S. and Ph.D. in EECS from U.C. Berkeley in 1990 and 1992, respectively. He was a professor in the E&CE Dept. of the University of Waterloo, Canada, from 1992 to 2000. Since April 2000, he has been a professor in both the CS&E and EE Depts. of the Pennsylvania State University. He has served as TPC co-chair of IEEE INFOCOM and is currently an IEEE Senior Member. His research on networking, including this, his third short book, has been generously supported by the National Science Foundation and Cisco Systems.

Printed in the United States
by Baker & Taylor Publisher Services